Disasters and Democracy

DISASTERS
AND
DEMOCRACY
The Politics of Extreme Natural Events

RUTHERFORD H. PLATT

WITH CONTRIBUTIONS BY

Miriam Gradie Anderson, Alexandra D. Dawson,
Jessica Spelke Jansujwicz, Ute J. Dymon,
K. Beth O'Donnell, Claire B. Rubin, and David Scherf

ISLAND PRESS
Washington, D.C. ◆ Covelo, California

Library of Congress Cataloging-in-Publication Data
Platt, Rutherford H.
    Disasters and democracy : the politics of extreme natural events /
Rutherford H. Platt with contributions from Miriam Gradie Anderson
. . . [et al.].
        p.   cm.
    Includes bibliographical references and index.
    ISBN 1-55963-696-3 (acid-free paper)
    1. Disaster relief—Political aspects—United States.   2. Disaster
relief—Law and legislation—United States.   3. Natural disaster—
Government policy—United States.   4. Natural disasters—Law and
legislation—United States.   I. Title.
    HV555.U6P53 1999   99-18910
    363.34'0973—dc21   CIP

Printed on recycled, acid-free paper

Manufactured in the United States of America
10 9 8 7 6 5 4 3

To Anne and Steve

ABOUT ISLAND PRESS

Island Press is the only nonprofit organization in the United States whose principal purpose is the publication of books on environmental issues and natural resource management. We provide solutions-oriented information to professionals, public officials, business and community leaders, and concerned citizens who are shaping responses to environmental problems.

In 1999, Island Press celebrates its fifteenth anniversary as the leading provider of timely and practical books that take a multidisciplinary approach to critical environmental concerns. Our growing list of titles reflects our commitment to bringing the best of an expanding body of literature to the environmental community throughout North America and the world.

Support for Island Press is provided by The Jenifer Altman Foundation, The Bullitt Foundation, The Mary Flagler Cary Charitable Trust, The Nathan Cummings Foundation, The Geraldine R. Dodge Foundation, The Charles Engelhard Foundation, The Ford Foundation, The Vira I. Heinz Endowment, The W. Alton Jones Foundation, The John D. and Catherine T. MacArthur Foundation, The Andrew W. Mellon Foundation, The Charles Stewart Mott Foundation, The Curtis and Edith Munson Foundation, The National Fish and Wildlife Foundation, The National Science Foundation, The New-Land Foundation, The David and Lucile Packard Foundation, The Pew Charitable Trusts, The Surdna Foundation, The Winslow Foundation, and individual donors.

CONTENTS

ACKNOWLEDGMENTS

The research on which this book is based was funded in part by National Science Foundation Grant No. 93-19422. I would like to thank Dr. Eleanora Sabadell, the NSF project officer, for her support and patience over the course of the study.

I would like to thank all of my collaborators: Ute Dymon, Claire Rubin, Alexandra Dawson, and former students Miriam Gradie Anderson, Jessica Spelke Jansujwicz, Beth O'Donnell, and David Scherf. Other former students who contributed to the study in various ways include James Kendra, Van Morrill, Steven Whitman, and Kenneth Holmberg. Frank Boscoe assisted Ute Dymon with the preparation of the maps in Chapter 2. Andrew Dolan prepared Figure 1-7.

Like all hazard researchers, I owe a debt of gratitude to the Natural Hazards Research and Applications Center at the University of Colorado-Boulder, in particular to its founder and mentor of us all, Gilbert F. White, and to its co-directors Mary Fran Myers and Dennis Mileti and their colleagues David Morton, David Butler, and Sylvia Dane.

Among many other professional colleagues who have contributed in one way or another, I would like to mention Jon Kusler, J. Kenneth Mitchell, Norbert Psuty, Raymond Burby, Peter May, David Godschalk, Orrin Pilkey, Richard Sylves, Susan Cutter, Frank Thomas, James Wright, Edward J. Blakely, and Kenneth Topping.

Many federal agency staff helped us with the research in Washington, D.C., and in the areas covered by the case studies. At FEMA headquarters, I would like to thank Richard Krimm, Michael Buckley, Mark Crowell, Michael Robinson, Edward Pasterick, Mark Whitney, Cliff Oliver, William Lesser, and H. Crane Miller, among others. In FEMA regional offices, I would like to thank Edward Thomas, Kevin Murley, and Michelle Steinberg (Reg. I), Joseph Picciano and Pat Griggs (Reg. II), Todd Davison (Reg. IV), David Schein (Reg. V), and Ron McCabe (Reg. VII). I also thank: Becky Brantley, Small Business Administration; Jan Opper, Department of Housing and Urban Development; Keith Bea, Congressional Research Service; David Wood, General Accounting Office; Alan Rhinesmith, Office of Management and Budget; Dan Pryor, Office of Science and Technology Policy; and Diane Abell, Fire Island National Seashore, National Park Service.

Innumerable people with state and local governments, environmental groups, and property rights organizations helped us in various way. Space does not allow them all to be acknowledged here, but many are referenced in the text.

I would also like to thank the loyal office staff of the Department of Geosciences at the University of Massachusetts, Amherst, including Martha Beckwith, Nicole Pietraszkiewicz, Ann Lucas, and Carol Vogel.

Finally, Kathleen Lafferty performed editing and general preparation of the manuscript with good humor and calm efficiency.

Rutherford H. Platt

PREFACE

And so, my fellow Americans, ask not what your country can do for you; ask what you can do for your country.

President John F. Kennedy,
Inaugural Address, January 20, 1961

Residents welcomed a two-star general from the Army Corps of Engineers here this week, taking him on a tour of their eroded beaches and urging him to support a $60 million plan to save their seashore.

"Engineer Corps Weighs Shoring Up Fire Island,"
The New York Times, June 30, 1996

Natural disasters in the United States, including floods, hurricanes, coastal erosion, tornadoes, wildfires, and earthquakes, on the average cause roughly $20 billion annually in direct costs to government, to the insurance industry, and to victims, and these costs are continuing to escalate.[1] In addition, disasters inflict unmeasurable noneconomic losses in terms of deaths, injuries, environmental and habitat damage, loss of treasured possessions and pets, disruption of family life, and emotional turmoil.[2]

Our nation's history is littered with disasters. The twentieth century opened with an epic but unnamed hurricane which killed 6000 people at Galveston, Texas, in 1900. (Some other historic catastrophes are listed in the Introduction, which follows.) But disaster as a topic of daily news coverage and instantaneous governmental response is a relatively recent phenomenon. So prominent has become the perception of natural disasters in contemporary society that the United Nations has designated the 1990s as the International Decade for Natural Disaster Reduction, and over 100 countries including the United States have established counterpart national programs to reduce the harm inflicted by extreme natural events.

Experts debate the extent to which the perceived rise in disaster losses is attributable to (1) increasing vulnerability due to encroachment of human activities on hazardous sites; (2) increasing risk from weather and sea-level-related hazards due to global climate change; or (3) increasing capabilities for reporting and documenting disaster losses (the "CNN syndrome"). Presumably, all three factors underlie, to an unknown extent, the rising incidence and awareness of natural disasters as a recurrent source of disruption and misery in the late twentieth century.

This book does not attempt to resolve this riddle. The physical and human causes of natural disaster are well documented by others.[3] But we are here concerned with political response to natural disasters, however they may arise. Specifically, we are interested in the changing role of the federal government from distant observer to immediate responder, principal financier of disaster costs, and, more recently, champion of hazard mitigation.

Few would cavil about the need for federal involvement in true catastrophes such as Hurricane Andrew or the Northridge Earthquake, or the provision of immediate aid to lower-income victims of disasters of any magnitude. But presidential declarations of "major disasters" are announced nearly every week, with 72 declarations issued in fiscal year 1996, 49 in 1997, and 68 in 1998.[4] Among the last, a declaration of "major disaster" following a heavy rainstorm in Boston, Massachusetts, in June 1998 inspired the following skeptical comment in a front page article in *The Boston Globe*:

> Be careful. It's a disaster out there. That's the word from President Clinton, [who] took time out from preparing for his trip to China to declare the city of Boston and all of Suffolk, Middlesex, Essex, Bristol, and Norfolk counties a national disaster area. . . . Many people expressed surprise at the disaster declaration with its dark connotations of ravage and ruin. No one wanted to return any federal relief dollars, but some people wondered at the ease with which the president can bestow federal funds on even a not-quite-reeling community.[5]

Clearly, something has changed in our political definition of, and reaction to, natural disasters in the United States.

In this book we are not principally concerned with the total number of disasters declared to be eligible for federal assistance, or with the total cost of federal disaster assistance programs per se. Rather, in common with numerous recent critiques of federal disaster policies, we raise the question as to whether federalizing the costs of disasters is helping to lighten the overall burden of disasters or, perversely, whether it is making matters worse. Has the availability of federal disaster assistance, flood insurance, and other benefits inadvertently contributed to a false sense of security, roughly equivalent to the discovery in the 1960s that federal flood control projects were stimulating development in "protected" floodplains? Those areas experienced greater losses when floods exceeded the design capacity of dams, levees, and sea walls than would have occurred without the projects.[6] Are people more likely to invest in property in hazardous locations in the belief that, if worse comes to worst, the federal government will hold them relatively harmless? Are communities, concerned about property values and tax revenue, more likely to allow, or even tacitly encourage, building or rebuilding in unsafe locations in the expectation of a federal bail-out if disaster strikes?

Closely related is another critical question: How much responsibility should individuals and local communities be expected to assume to protect themselves? A precept running through federal disaster legislation has been the need for states and local governments to curtail unwise development and redevelopment through land use planning and regulation. In the 1990s, however, that mandate has become muted as property rights interests have challenged the power of government to restrict the use of private land without compensation.

A paradox or double standard thus emerges. On the one hand, the federal government is called upon to assume a major share of state, local, and private economic costs of disasters through grants, subsidized loans, and government-backed insurance programs. But on the other hand, the government at all levels is increasingly impotent to demand as a condition of such benefits that local governments and individuals assume the political and financial burdens of curtailing unwise development in hazardous locations. "Give us the money, and no strings attached"—the political refrain of governors, mayors, and private investors—is duly chorused by members of Congress when disaster strikes their constituencies. The Federal Emergency Management Agency, as a creature of Congress, is thereby undercut in its efforts to reduce hazard vulnerability unless it antes up still more money to pay for hazard mitigation.

The research underlying this book was funded by National Science Foundation Award No. 9319422. Its scope was originally limited to the coast as reflected in the project title: "Federalism and the Coast: Loss Bearing versus Loss Avoidance." With the helpful approval of NSF staff officer Dr. Eleanora Sabadell, the study was expanded to encompass noncoastal disasters as well.

The Introduction and Chapter 1 trace the historical evolution of the federal disaster assistance role and major programs from virtually zero earlier in this century to a wide array of authorities involving some 26 federal agencies today. Chapter 2 summarizes a spatial analysis by Dr. Ute Dymon, a geographer at Kent State University, of disaster declarations and federal assistance provided under the Stafford Act since its adoption in 1988. In Chapter 3, Claire B. Rubin and I jointly examine the concept of hazard mitigation, historically and in present practice. While mitigation has been widely publicized under the Clinton administration, a major shortcoming has been the gradual erosion of earlier congressional mandates that land use regulation is the "key" to reducing vulnerability to flood hazards, the major source of national disaster losses.

Looking behind the "backlash against regulation," Jessica Spelke Jansujwicz in Chapter 4 examines the property rights movement as of the mid-1990s, with profiles of several national and regional property rights organizations. This leads into a detailed review of constitutional law on the "taking issue" as it applies to the regulation of hazardous areas in Chapter 5, written by attorney Alexandra D. Dawson and myself. We argue that the claims of the property rights advocates, based on *Lucas v. South Carolina Coastal Council* are overblown: government still has both the power and the responsibility to use its customary regulatory power (the "police power") to reduce vulnerability to life and property.

Chapters 6, 7, and 8 comprise major regional case studies that examine the federal role and the balance of federal and nonfederal loss-bearing in three disaster recovery contexts: Fire Island, New York, after the 1992–93 winter storms (Chapter 6); St. Charles County, Missouri, after the Midwest Flood of 1993 (Chapter 7), and Oakland, California, after the Loma Prieta Earthquake of 1989 and the urban-wildland fire of October 20, 1991 (Chapter 8). Two of these studies were prepared in collaboration with former students David Scherf and Beth O'Donnell (Chapter 6) and Miriam Anderson (Chapter 7).

These narratives collectively validate some of the critiques and issues reviewed earlier in the book. In particular, the hypothesis is verified that national disaster policy is increasingly skewed toward federal reimbursement of disaster costs and away from individual and local responsibility for loss avoidance, except on a voluntary basis. Mitigation has become a dominant theme in the disaster response vocabulary of the 1990s, but it has also become increasingly understood to mean whatever the federal government pays for, rather than what local governments and individuals can be expected to do to protect themselves without compensation.

Cross-cutting issues and recommendations for federal policy are summarized in the Conclusion. The book's major recommendations are: (1) depoliticize FEMA individual assistance by removing it from the presidential declaration process and transferring it to states or the American Red Cross; (2) reduce the magnitude and cost-share of FEMA public assistance through greater reliance on loans, insurance, and community needs-based criteria; (3) eliminate federal incentives to build or rebuild in areas subject to repetitive natural hazards; and (4) strengthen nonfederal hazard mitigation through more effective use of land use controls and incentives.

NOTES

1. National Research Council, *A Safer Future: Reducing the Impacts of Natural Disasters.* Washington, D.C.: National Academy Press (1991) p. 7. See also: William J. Petak and Arthur A. Atkisson, *Natural Hazard Risk Assessment and Public Policy: Anticipating the*

Unexpected. New York: Springer-Verlag (1982). This study estimated that costs from nine types of natural disasters in the United States would rise from $17.1 billion in 1970 to $37.7 billion in 2000 (in 1980 constant dollars). Also see: Insurance Institute for Property Loss Reduction, *Coastal Exposure and Community Protection: Hurricane Andrew's Legacy.* Boston: IIPLR (1995), which estimates that private insurance coverage against natural disasters in counties bordering the U.S. Atlantic and Gulf of Mexico coasts has increased by 69 percent from $1.86 trillion to $3.15 trillion between 1988 and 1993. One broader estimate by the Federal Office of Science and Technology Policy places total U.S. disasters costs at $1 billion *per week* over the 5-year period from 1992 to 1996. (unpublished data). See also a report on "Hidden Costs of Coastal Hazards: Implications for Risk Assessment and Mitigation," in preparation by the H. John Heinz III Center for Science, Economics, and the Environment, Washington, D.C. A related report is also in preparation by the National Research Council Committee on Assessing the Costs of Natural Disasters.

2. There is a vast literature on the costs of natural disasters in the United States and elsewhere. See, for example, Walter Peacock, Betty Hearn Morrow, and Hugh Gladwin, eds., *Hurricane Andrew: Ethnicity, Gender, and the Sociology of Disaster.* London: Routledge (1997); Dennis Mileti (In press). *Disasters by Design: A Reassessment of Natural Hazards in the United States.* Washington, D.C.: Joseph Henry Press; Raymond J. Burby et al., *Sharing Environmental Risks: How to Control Governments' Losses in Natural Disasters.* Boulder, Colo.: Westview Press (1991); Frederick C. Cuny, *Disasters and Prevention.* New York: Oxford University Press (1983); Louise K. Comfort, ed., *Managing Disaster: Strategies and Policy Perspectives.* Durham, N.C.: Duke University Press (1988); Pamela S. Showalter, William E. Reibsame, and Mary Fran Myers, *Natural Hazard Trends in the United States: A Preliminary Review for the 1990s.* Working Paper No. 83. Boulder, Colo.: Institute of Behavioral Science; National Research Council, *Facing the Challenge. U. S. National Report to the IDNDR* (1994).

3. For example: Ian Burton, Robert W. Kates, and Gilbert F. White, *The Environment as Hazard* (2nd ed.). New York: The Guilford Press (1993); David Alexander, *Natural Disasters.* London: Chapman and Hall (1995); Keith Smith, *Environmental Hazards: Assessing Risk and Reducing Disaster* (2nd ed.). London: Routledge (1996); Andrew Kirby, ed., *Nothing to Fear: Risks and Hazards in American Society.* Tucson: University of Arizona Press (1990); Risa Palm, *Natural Hazards: An Integrative Framework for Research and Planning.* Baltimore: Johns Hopkins University Press (1990); E.L. Quarantelli, ed., *What Is a Disaster? Perspectives on the Question.* London: Routledge (1998); Susan Cutter, ed., *Environmental Risks and Hazards.* Englewood, N.J.: Prentice-Hall (1994). See further titles in Selected Bibliography.

4. Declarations apply to individual states, so multiple declarations may relate to a single disaster affecting more than one state.

5. Stephen Kiehl, "Little Notice for Local 'Disaster.'" *The Boston Globe* (June 25, 1998): pp. 1 and 33.

6. Gilbert F. White, *Choice of Adjustment to Floods.* Research Paper No. 93. Chicago: University of Chicago Department of Geography (1964).

DISASTERS AND DEMOCRACY

Introduction

DISASTERS BEFORE 1950: COPING WITHOUT CONGRESS

For its first 160 years of nationhood, the United States had no general policy or program for responding to natural or human-caused disasters. Such catastrophes as the New Madrid, Missouri, Earthquakes of 1811–1812, the Chicago Fire of 1873, the Johnstown, Pennsylvania, Dam Break in 1889, the Galveston Hurricane of 1900, the San Francisco Earthquake and Fire of 1906, the Miami Hurricane of 1926, the Lower Mississippi Flood of 1927, and the New England Hurricane of 1938 ravaged portions of the nation periodically. Deaths from such disasters numbered in the hundreds, and sometimes in the thousands. Costs in present-day terms ran into the billions of dollars. The costliest hurricane in American history struck Miami in 1926 causing an estimated $39 billion in damage as adjusted to 1992 dollars, considerably higher than the $24 billion in losses attributable to Hurricane Andrew in 1992.[1] Yet while the latter event triggered an avalanche of federal assistance, the 1926 storm went almost unnoticed nationally.

The U.S. Congress was not entirely unmoved by these and other disasters: Between 1803 and 1947, 128 specific acts expressed sympathy and sometimes supplied token financial assistance.[2] But the tasks of actual response, rescue, repair, and reconstruction were organized under local auspices, with financial and other assistance contributed by states, cities, churches, and various sources outside the disaster area. Furthermore, reduction of vulnerability to natural hazards ("mitigation"), was accomplished, if at all, through actions taken individually or at the local level prior to the 1930s. In 1888 when the Brighton Beach Hotel at Coney Island, New York, was threatened by erosion, its owner—the Brooklyn, Flatbush, and Coney Island Railroad—jacked the structure up and hauled it 450 feet landward on a flotilla of flatcars.[3] After a hurricane killed over 6000 people at Galveston, Texas, on September 8, 1900, homes and streets were gradually elevated and Galveston County built a six-mile, 16-foot-high sea wall.[4] The Army Corps of Engineers initially provided only technical assistance, although it later extended the wall another four miles.[5] After the 1906 San Francisco Earthquake, the nation's greatest urban disaster thus far, the federal role was chiefly to provide Army troops to deter looting. The American Red Cross,

1

working closely with other charitable and civic organizations, provided food and shelter using funds and supplies sent from communities across the nation and from several foreign countries.[6]

When the Lower Mississippi River broke through its levees in 1927, spreading across 20,000 square miles of floodplains, governors of six affected states pleaded for federal assistance. U.S. Secretary of Commerce Herbert Hoover assumed direction of emergency response involving state and local authorities and their militias, the Corps of Engineers, the Coast Guard, a naval air contingent, the Weather Bureau, and the Red Cross.[7] They had their hands full: 162,000 homes were flooded, 41,000 of them destroyed, and 325,000 people were cared for in Red Cross camps and 311,000 others fed by the Red Cross in private homes.[8] While financial assistance to victims was largely provided through private contributions, that disaster stimulated federal response of a different nature, namely measures to control flood waters through the construction of dams, levees, and diversion channels. The Lower Mississippi Flood Control Act of 1928 and its successors in 1936 and 1938 launched the Corps of Engineers multibillion dollar program to tame the nation's major rivers. Referring to the 1927 flood, hydrologists William G. Hoyt and Walter B. Langbein have written: "Few natural events have had a more lasting impact on our engineering concepts, economic thought, and political policy in the field of floods. Prior to 1927 control of floods in the United States was considered largely a local responsibility. Soon after 1927, the control of floods became a national problem and a federal responsibility."[9] (See Figure I-1.)

While the federal government thus assumed increasing responsibility to build flood control projects as part of the larger New Deal public works program to combat the Great Depression, it remained aloof from providing direct assistance to disaster victims. Before 1950, disaster assistance was viewed as the moral responsibility of neighbors, churches, charities, and communities—not the federal government. Furthermore, disasters tended to be viewed as unavoidable "acts of God," which, by definition, transcend the power of government to prevent. According to a 1916 court decision: "Acts of God . . . are those events, accidents, or manifestations of nature which proceed from natural causes, and which are unusual and unprecedented in character and cannot be reasonably anticipated or guarded against by the exercise of ordinary care."[10] The availability of this common law defense, while not always sufficient to protect a defendant from liability, served to cloak government inaction with a pseudo-theological alibi: "What will be, will be." It is outmoded today in light of improved scientific understanding of place-specific natural hazards.[11]

Fatalism regarding natural disasters was reinforced by the prevailing gospel of "laissez-faire," namely that government should never interfere in the private market economy. This doctrine originated with the eighteenth-century English

FIGURE I-1 Sandbagging along the Connecticut River, 1927. (Photo: Holyoke Water Power Co.)

political philosopher Adam Smith, who wrote in *The Wealth of Nations* that an individual who "intends only his own gain" is "led by an invisible hand to promote . . . the public interest."[12] The doctrine of laissez-faire stifled attempts to regulate corporate actions or building practices even when human negligence and avarice were manifest. For instance, after the dam failure that killed 2209 people in Johnstown, Pennsylvania, on May 31, 1889, government at all levels was silent despite widespread public outcry in the press. The earthen dam and lake that it impounded were the private property of a Pittsburgh millionaire club whose members used the facility as a summer retreat. Despite ample warnings that the dam was unsafe, no sufficient repairs were made. The resulting disaster was judicially deemed an "act of God" and the club was never held responsible.[13] Congress provided no disaster relief and did nothing to ensure that such a disaster would not be repeated there or elsewhere. This was the golden age of laissez-faire and deference to wealth.

Cities were capable of addressing their public infrastructure needs, but usually allowed private land to be built or rebuilt as the market saw fit. San Francisco after its 1906 earthquake and fire acted boldly to expand its water supply, whose failure was blamed for the burning of much of the city. With approval of the federal government (but over the objections of John Muir), the city dammed the Hetch Hetchy River in Yosemite National Park and built an aqueduct to lead water 150 miles from the Sierra Nevada to the city. But in the rush to rebuild the city itself, private capital insisted on following the original grid street pattern, ignoring a new city plan for redevelopment more compatible with the city's unique and hazardous site.[14] (See Chapter 8.)

By the early 1900s, the progressive reform movement was beginning to challenge laissez-faire as outdated and dangerous in a complex capitalist society. Among the many concerns of reformers such as Lincoln Steffens, Jacob Riis, and Theodore Roosevelt were corporate monopolies, waste of natural resources, working conditions in factories, banking reform, housing for the poor, and urban congestion. Due in part to the immense popularity of Teddy Roosevelt, who ran for reelection to the presidency in 1912 on a third-party progressive platform (the so-called "Bull Moose party"), progressive reform proposals attracted strong support from mainstream middle class Americans during the opening decades of the century.[15]

Two shocking disasters of human origin focused reform attention specifically on the need for public intervention to protect lives from corporate misfeasance. On March 25, 1911, 146 young women were killed in a fire that ravaged a 10-story building housing the Triangle Shirt Waist factory in New York City. The building lacked sprinklers and fire escapes, and its upper floors were beyond the reach of fire fighting equipment. The tragedy helped to advance support for public laws to promote fire safety and limits on the heights of buildings.[16]

The sinking of *Titanic* on April 15, 1912, upon ramming an iceberg further inflamed public opinion against corporate arrogance. *Titanic* on its maiden voyage was attempting to break the speed record across the North Atlantic despite radio warnings of ice. The ship was vaunted to be "unsinkable" by the White Star Line, the British shipping company that built her. White Star had equipped *Titanic* with enough lifeboats and rafts to accommodate only about 1100 persons, or less than half of the number of individuals on board. A total of 1635 persons died in the catastrophe and merely 711 survived despite calm sea conditions and good visibility. Although legends of heroism abounded, a higher percentage of first class passengers and crew members survived than of lower class passengers. (White Star's senior representative on board was among the survivors although the captain was not.) This epic disaster called widespread attention in industrialized nations to the hazards of allowing private corporations excessive latitude where human life is at stake. It eventually led to strengthened international requirements for safety at sea, and paved the way for other forms of government oversight of hazardous activities.[17] (See Figure I-2.)

Decisions of the U.S. Supreme Court in the early years of the twentieth century displayed a conspicuous, although not entirely consistent, evolution from adherence to laissez-faire and Social Darwinism ("survival of the fittest") toward grudging acceptance of the need for governmental intervention in the private

FIGURE I-2 Steamship *Titanic* in Southampton, England, before her tragic voyage in April 1912. (Photo: The Titanic Historical Society, Indian Orchard, Massachusetts)

economy to protect the public health, safety, and welfare. The Court's 1905 decision in *Lochner v. New York*[18]—a dying gasp of the old order—rejected by a 5-4 vote a state law that regulated the hours of bakery employees as, in the Court's view, an unjustified interference with private property. But where private actions posed a danger to the public at large, progressive legal scholars urged the need for broader public powers to prohibit such harmful conduct. The year before *Lochner*, Professor Ernst Freund of the University of Chicago Law School in his seminal treatise, *The Police Power*, argued that common law nuisance, a doctrine derived from medieval England, was ill-suited to managing the perils of modern technology and urban society. In Freund's view, reasonable government regulation of private activity—the "police power"—was needed to safeguard the public:

> The common law of nuisance deals with nearly all the more serious and flagrant violations of the interests which the police power protects, but it deals with evils only after they have come into existence, and it leaves the determination of what is evil very largely to the particular circumstance of each case. The police power endeavors to prevent evil by checking the tendency toward

it and it seeks to place a margin of safety between that which is permitted and that which is sure to lead to injury or loss. This can be accomplished to some extent by establishing positive standards and limitations which must be observed, although to step beyond them would not necessarily create a nuisance at common law.[19]

Perhaps swayed by Freund's rationale, the Supreme Court four years after *Lochner* upheld a Massachusetts law imposing height limits on new buildings in downtown Boston for reasons of fire safety.[20] In 1915, it approved an ordinance of the city of Los Angeles that closed down existing brickyards in newly settled portions of the city as a public health measure.[21]

The evolution of judicial progressivism with respect to public intervention in the use of private land culminated in the 1926 decision in *Ambler Realty v. Village of Euclid,*[22] which upheld the practice of urban land use zoning (since referred to by planners as "Euclidean zoning"). The First National Conference on City Planning and Congestion in 1909 had aroused interest in planning and zoning among urban reformers. In 1916, the first comprehensive zoning law was adopted by New York City; a decade later it had spread to several hundred cities.[23] Coincidentally, the 1920 Census recorded more urban than rural residents for the first time in the nation's history. The majority opinion in *Euclid* acknowledged that recent urbanization of the nation called for fresh thinking about the permissible balance of private and public rights under the Constitution: "Building laws are of modern origin. They began in this country about 25 years ago. Until recent years, urban life was comparatively simple; but with the great increase and concentration of population, problems have developed . . . which require . . . additional restrictions in respect of the use and occupation of private lands in urban communities. Regulations, the wisdom, necessity and validity of which as applied to existing conditions, are so apparent that they are now uniformly sustained under the complex conditions of our day. . . ."[24] (See Chapter 5).

From the 1920s onward, Euclidean zoning was widely employed to discourage "discordant" patterns of urban land use—to keep apartment buildings out of single family neighborhoods or adult bookstores away from elementary schools. But ironically it was rarely used to restrict building in areas vulnerable to natural hazards. While the nation spent huge sums to store, divert, and channelize floodwaters, land use regulations were seldom used to limit new development in areas of flood risk. Federal flood control projects in fact often made matters worse by providing a false sense of security that attracted new development to floodplains. Floods exceeding the design limits of projects inflicted greater annualized damage than would have occurred in the absence of the project.[25] Until 1950, however, the federal government assumed no responsibility for the

FIGURE I-3 Hartford, Connecticut, during the flood of March 1936. (Photo: Army Corps of Engineers)

personal and economic welfare of flood victims, even when their plight was attributable to governmental assurances regarding the reliability of flood control projects.

Nor did Congress perceive any need for governmental restrictions on new encroachments in floodplains (or other areas of natural hazard for that matter). In 1955, Hoyt and Langbein observed that "Flood zoning, like almost all that is virtuous, has great verbal support, but almost nothing has been done about it. A few local governments have restricted the use of low-lying lands, but not enough for us to point to any substantial amount of experience, or any great degree of progress."[26] (See Figure I-3.)

The concept of floodplain management through local land use planning zoning would finally be embraced by Congress in the National Flood Insurance Act of 1968, but it remains controversial to the present time. The ghost of Adam Smith still haunts efforts at all levels of government to deter private investment in hazardous locations. However, as the costs of improvident land use decisions have proliferated, the nation's willingness to underwrite a substantial portion of those costs has been radically transformed by Congress and an increasingly compliant White House.

NOTES

1. Roger A. Pielke, Jr. (1997). "Normalized Hurricane Damages in the United States: 1925–1995." (Draft article submitted to *Weather and Forecasting*). Pielke uses a combination of inflation, wealth, and housing density changes to normalize damage totals.

2. Peter J. May (1985). *Recovering from Catastrophes: Federal Disaster Relief Policy and Politics.* Westport, Conn.: Greenwood Press, 20.

3. *Scientific American* (1888). "Moving the Brighton Beach Hotel" (April 14): 1–2; Rutherford H. Platt and others (1992). *Coastal Erosion: Has Retreat Sounded?* Program on Environment and Behavior Monograph No. 53. Boulder: Institute of Behavioral Science, University of Colorado, x.

4. National Research Council (1987). *Responding to Changes in Sea Level.* Washington, D.C.: National Academy Press, 82–83.

5. Martin Reuss (1991). U.S. Army Corps of Engineers Office of History, Personal Communication.

6. William Bronson (1959/1989). *The Earth Shook, The Sky Burned.* San Francisco: Chronicle Books, 100–104.

7. William G. Hoyt and Walter B. Langbein (1955). *Floods.* Princeton: Princeton University Press, 263.

8. Pete Daniel (1977). *Deep'n As It Come: The 1927 Mississippi River Flood.* New York: Oxford University Press, 10.

9. Hoyt and Langbein, note 7, 262–263.

10. David Alexander (1993). *Natural Disasters.* New York: Chapman and Hall, 342.

11. Ibid.

12. Adam Smith (1776/1937). *The Wealth of Nations.* New York: Modern Library, 423.

13. David G. McCullough (1968). *The Johnstown Flood.* New York: Simon and Schuster. McCullough reports that Johnstown received an immense outpouring of donated money, food, and building materials from around the United States and abroad.

14. Bronson, note 6, 174.

15. Richard Hofstader (1955). *The Age of Reform.* New York: Knopf.

16. Seymour Toll (1969). *Zoned American.* New York: Grossman Publishers, 26.

17. Wynn Craig Wade (1979). *Titanic: End of the Dream.* New York: Penguin.

18. 198 U.S. 45 (1905).

19. Ernst Freund (1904). *The Police Power: Public Policy and Constitutional Law.* Chicago: Callaghan.

20. *Welch v. Swasey* 214 U.S. 91 (1909).

21. *Hadachek v. Sebastian* 239 U.S. 304 (1915).

22. 272 U.S. 365 (1926).

23. Rutherford H. Platt (1996). *Land Use and Society: Geography, Law, and Public Policy.* Washington, D.C.: Island Press, Ch. 7.

24. 272 U.S., 386–387.

25. Gilbert F. White (1960/1986). "Strategic Aspects of Urban Floodplain Occupance," in *Geography, Resources, and Environment* Vol. 2 (eds. Robert W. Kates and Ian Burton). Chicago: University of Chicago Press, 84–96.

26. Hoyt and Langbein, note 7, 95.

PART I

FEDERALIZING DISASTERS:
FROM COMPASSION TO ENTITLEMENT

The Federal Disaster Relief Act of 1950 marked the beginning of a half-century of federal laws, programs, and policies intended to soften the financial and social impacts of natural disasters on the American people and their communities. Chapter 1 traces the evolution of such federal initiatives in terms of purpose (e.g., individual relief, community reconstruction, hazard mitigation) and means (e.g., disaster assistance grants, subsidized disaster loans, federal insurance programs, technical assistance). The chapter concludes with the problem—widely raised by disaster policy critiques since the mid-1990s—of "moral hazard," namely, to what extent does the likelihood of generous federal assistance serve to diminish the natural caution that individuals, communities, and businesses might otherwise exercise in adjusting to natural hazards in their investment and locational decisions? At what point does compassion lead to "codependency" whereby potential disaster victims and their federal protectors become locked into a repetitive cycle of loss, compensation, reconstruction, and new losses?

The spatial incidence of disasters, and of disaster assistance, is by no means uniform. Some places—notably coastal California, the Gulf Coast, and parts of the Mississippi River Valley—seem to be more disaster-prone and therefore receive greater allocations of federal assistance than "safer" places. Federal disaster programs are so diverse in form and function that mapping their collective spatial impact would be a daunting task indeed. However, Chapter 2 examines

FEDERAL DISASTER ASSISTANCE

The nation's first general disaster assistance law—the Disaster Relief Act of 1950 (PL 81-875)—was a mere pinprick in relation to this backdrop of federal activism and grandiose construction programs. Three months after the outbreak of the Korean War and with the foul winds of McCarthyism beginning to waft through the nation's capital, it was adopted quietly with little study or debate. Its initial authorization was $5 million, paltry even then. But despite the conservative climate of the moment, it harked back to New Deal social legislation as a logical extension of social security, housing, education, VA medical care, and other social benefit programs. Initially, its benefits were limited to local public costs; later this would be expanded to include private enterprise and individuals as well. The 1950 law marked a threshold in national policy concerning disasters, from an era of disinterest to one of limited federal involvement (Table 1-1). It was the modest forerunner of a long series of acts that would cumulatively commit the United States to providing tens of billions of dollars in assistance to individuals and communities stricken by natural and other disasters.

This obscure law was introduced by Rep. Harold Hagen of Minnesota, whose immediate concern was to relieve the financial burdens of repairing the farm-to-market roads and bridges in recently flooded areas along the Red River in Minnesota and North Dakota (the same region where prolonged floods would attract national attention and $2 billion in federal assistance in 1997). Its sponsor presented a long list of past special acts of Congress, along with other examples of federal response to disasters through the Army Corps of Engineers, the Farm Credit Administration, and the Bureau of Public Roads. But unlike those ad hoc measures, the new legislation would prove to be the first permanent and general disaster law passed by Congress, and its concepts would become the model of all succeeding federal disaster laws, albeit vastly expanded in scope and cost.[1] Its sketchy legislative history states that the law: "will cover disaster occurring anywhere in the entire country instead of a particular state or locality. [It will] provide for an orderly and continuing method of rendering assistance to the states and local governments in alleviating suffering and damage resulting from a major peacetime disaster. . . ."[2] The means provided to accomplish that mission were, however, parsimonious as compared with the disaster cornucopia of the 1990s. The several hurricanes that struck the East Coast during the mid-1950s introduced the public to the new concept of federal disaster assistance, albeit in limited form (Figure 1-1).

The peacetime mission of the fledgling program was to be overshadowed for decades by the Cold War. It was initially assigned to the Housing and Home Finance Agency, which administered the federal urban renewal program. But from 1953 until 1974, the program was housed within a series of civil defense agencies where it languished in relative obscurity in the midst of preparations for nuclear war (Table 1-2). In 1974, it returned to the arena of community plan-

TABLE 1-1 Disasters and Congress: A Selective Chronology

Early Period—Negligible or Ad Hoc Federal Assistance

1889	Johnstown, Pennsylvania, dam break: 2209 deaths
1900	Galveston, Texas, hurricane: over 6000 deaths
1906	San Francisco Earthquake and Fire
1926	South Florida Hurricane
1927	Lower Mississippi River Flood
1928	Lower Mississippi Flood Control Act of 1928
	St. Francis Dam (Calif.) break: over 400 deaths
1933	Long Beach Earthquake (California)
1936	Floods: Ohio and Lower Mississippi River Basins
1938	New England Hurricane
1944	South Florida Hurricane

Transitional Period—Limited Federal Disaster Assistance

1950	Disaster Relief Act of 1950 (PL 81-875)
1953	Tennessee Valley Authority Local Flood Regulation Program established
	Small Business Administration Disaster Loan Program
1954-1955	New England Hurricanes
1956	National Flood Insurance Program Act (79 Stat. 1978)
1964	Alaskan Earthquake
	Alaskan Earthquake Assistance Act (PL 88-451)
1965	Hurricane Betsy: Gulf of Mexico
	Southeast Hurricane Disaster Relief Act (PL 89-339)
1968	National Flood Insurance Act (PL 90-448, Title XIII)
1969	Hurricane Camille: Gulf of Mexico and inland areas
	Disaster Relief Act (PL 91-79)
1970	Disaster Relief Act (PL 91-606)
1971	San Fernando Earthquake (California)
1972	Tropical Storm Agnes (Middle Atlantic states)
	Rapid City flash flood (South Dakota)
	Federal Dam Safety Act (PL 92-367)
1973	Flood Disaster Protection Act (PL 93-234)
1974	Disaster Relief Act (PL 93-288)
	Water Resources Development Act (PL 93-251)
1977	Earthquake Hazards Reduction Act (PL 95-124)
	Executive Order 11988: nonstructural floodplain management
1979	FEMA established
	Hurricane Frederic (Gulf of Mexico coast)
1980	Hurricane David (Gulf of Mexico coast)
1982	Coastal Barrier Resources Act (PL 97-348)

Recent Period—Abundant Federal Disaster Assistance

1988	Robert T. Stafford Disaster Relief and Emergency Assist. Act (PL 100-707)
1989	Hurricane Hugo (Caribbean and Southeast Atlantic states)
	Loma Prieta Earthquake (California)
1991	Oakland, California, wildfires

(continues)

TABLE 1-1 Continued

1992	Hurricane Andrew (Florida and Louisiana)
	Hurricane Iniki (Hawaii)
1993	Midwest Floods
	Hazard Mitigation and Relocation Assistance Act (PL 103-181)
1994	Northridge Earthquake (California)
	National Flood Insurance Reform Act (PL 103-325)
1995–1997	California Floods
1996	Hurricanes Fran and Bertha (North Carolina)
1997	Ohio River Floods
	Red River Floods (Minnesota, North Dakota)

FIGURE 1-1 Destruction at Winsted, Connecticut, due to flooding of the Mad River in Hurricane Diane, 1955. (Photo: Connecticut River Valley Flood Control Commission)

TABLE 1-2 Federal Disaster Agencies 1951 to Present

Approx. Dates	Responsible Agency
1951–1952	Housing and Home Finance Agency (HHFA)
1953–1958	Federal Civil Defense Administration (FCDA)
1958–1962	Office of Civil and Defense Mobilization (OCDM)
1962–1974	Office of Emergency Planning (OEP)
1974–1979	Federal Disaster Assistance Administration (FDAA) of U.S. Dept. of Housing and Urban Development (HUD)
1979–Present	Federal Emergency Management Agency (FEMA)

Source: Adapted from Frank P. Bourgin (no date), "Legislative History of Federal Disaster Relief, 1950–1974." Washington, D.C.: FEMA (mimeo).

ning under the newly created Federal Disaster Assistance Administration of the U.S. Department of Housing and Urban Development. From there, in 1979, it was transferred to the new Federal Emergency Management Agency, an amalgam of civilian and military preparedness programs established by President Jimmy Carter. In the 1980s, under the Reagan Administration, FEMA pursued a chimera called "integrated emergency management" that was intended to protect the American people from anything between a local flood and all-out nuclear war. Since 1993, under the Clinton Administration, FEMA has finally shed its Cold War baggage and has evolved into a genuinely domestic program with a strong emphasis on natural hazard mitigation. (See Chapter 3.)

After 1950, and more emphatically after the Federal Disaster Relief Act of 1970, the federal government assumed a permanent role as the primary source of funds and expertise to deal with major and some not-so-major disasters. The present dominance of federal assistance in disaster recovery was not intended or foreseen in the original 1950 law. As originally established by Congress, the federal disaster assistance program was to be:

1. Limited as to the scope of federal assistance to be supplied
2. Contingent upon a presidential disaster declaration finding that federal assistance is required to supplement state and local capabilities
3. Limited as to amounts of federal funding to be allocated to disaster relief

As discussed in following sections, the first of these limitations—*scope* of assistance—was gradually modified by amendments to the act toward a much broader range of benefits. The second—that presidential declarations should assure that federal assistance is *supplementary*—still receives lip service in the current version of the disaster assistance law (the 1988 Stafford Act), but has been found to be empty rhetoric by various policy reviews. Third, the *cost* of disaster assistance has risen from the initial appropriation of $5 million in 1950 to levels as high as $10 billion in 1994. Each of these limitations is considered in the following sections.

Scope of Assistance

At first, federal assistance was narrowly limited to emergency assistance and repairs to local public infrastructure. The 1950 act authorized only (1) the "utilization or loan" to states and local governments of federal equipment, supplies, facilities, personnel, and other resources; (2) distribution of food and medicine through the American Red Cross; (3) donation of surplus federal property to states and local governments; and (4) "performing on public or private lands protective and other work essential for the preservation of life and property, clearing debris and wreckage, making *emergency repairs to and temporary replacements of public facilities of local governments* damaged or destroyed in such major disaster . . ."[3] (emphasis added).

The rapid expansion in the scope of allowable benefits through 1974 has been documented by Mileti[4] and through 1980 by May.[5] In 14 acts between 1950 and 1980, the U.S. Congress vastly enlarged the disaster assistance functions of the federal government. Some of these laws responded to a specific disaster (e.g., the 1964 Alaskan Earthquake Assistance Act) while others amended or replaced the generic 1950 act (e.g., the 1974 Federal Disaster Relief Act). (See Table 1-1 for a partial list.) Among the benefits added to the basic federal disaster program during this period were temporary housing, grants for repair of damaged state property, unemployment compensation to disaster victims, legal and mental health services, individual and family grants, food coupons, and payments to communities to offset lost tax revenue.

A watershed in the federal role was crossed around 1969 with the authorization of benefits to individuals. Prior to that time, federal assistance was directed primarily to public units of government.

> It was not until the impact of Hurricane Camille on the Gulf Coast and Virginia and West Virginia that the Congress decided that a more formalized program of assistance to individuals should be instituted. Part of this was accomplished on October 1, 1969, with the passage of PL 91-79. [Additional acts passed in 1970 and 1974] continued and expanded this institutionalization of disaster assistance for individuals.[6]

Today, the scope of both individual assistance (IA) and public assistance (PA) is very broad. IA includes temporary housing, individual and family grants, unemployment compensation, food coupons, crisis counseling, and legal services. PA covers debris removal, repair, restoration, or replacement of public facilities of many types (including beaches and trees in certain cases), community disaster loans to cover shortfalls in local tax revenue due to a disaster (subject to cancellation),[7] and emergency response costs of states and local governments. All of the foregoing are subject to a 25 percent nonfederal cost share, unless it is reduced or waived by the president. (Related programs such as the National Flood Insurance Program and the Small Business Administration Disaster Loan Program are considered later in this chapter.)

IA and PA programs differ significantly with respect to eligibility. While both are contingent upon a presidential declaration and limited to counties thereby listed, most individual assistance is further limited by financial need criteria. Individual and family grants (IFGs), which are currently capped at about $14,000 per household, require that the recipients be too poor to qualify for a low-interest disaster loan from the Small Business Administration or other federal assistance. Temporary housing assistance requires that the applicant's primary residence has become unlivable and any insurance benefits available to the applicant are deducted from the federal grant.

By contrast, public assistance has no means test. Communities within declared counties are eligible for federal reimbursement of approved disaster-related costs to restore or replace damaged public facilities, *regardless of the economic status of the community or its residents*. Thus, financially hard-pressed individuals who do not meet the means criterion, as well as poor victims of nondeclared disasters, are ineligible for federal individual assistance. But affluent communities covered by declarations may receive between 75 percent and 100 percent of their recovery costs from the federal government, even if they carry or could afford to carry disaster insurance. As argued in the Conclusion, this inequity between IA and PA deserves reconsideration.

Presidential Declarations: The "Supplemental" Myth

Throughout the history of the federal disaster relief program, it has been steadfastly maintained in statute, regulation, and agency dogma that federal assistance is *supplementary* to state, local, and private resources. The legislative history to the 1950 Act clearly stated that limitation.

> The purpose of the bill is to provide for an orderly and continuing method of rendering assistance to the state and local governments in alleviating suffering and damage resulting from a major peacetime disaster and in restoring public facilities and *in supplementing whatever aid the state or local governments can render themselves*[8] (emphasis added).

Federal disaster assistance laws and regulations have never explicated how to assess what "aid the state or local governments can render themselves." The key action that unleashes federal benefits to stricken areas is a disaster declaration by

FIGURE 1-2 The Pearl River at flood stage in Jackson, Mississippi, April 1979. Flooded buildings in foreground were constructed in reliance on a new flood control levee (beyond the highway in distance) that failed. (Photo by R.H. Platt)

Congress and local governments and the proposal was dropped from further consideration. In 1988, the Stafford Act prohibited the use of an "arithmetic formula or sliding scale" to deny disaster assistance to any "geographic area."[20] In other words, Congress likes to keep the process imprecise, even if benefits occasionally go to the undeserving. The absence of objective criteria preserves wide political discretion to the president, a circumstance that has been questioned by some policy analysts such as Richard T. Sylves.[21]

Meanwhile the costs of disasters have been rising dramatically. Between 1970 and 1981, 376 major disasters and 84 emergencies were declared, leading to direct federal disaster assistance costs of $3.8 billion (unadjusted for inflation) for that period.[22] By contrast, between 1989 and 1994, 291 declarations cost the U.S. Treasury over $34 billion (unadjusted for inflation).[23]

In 1992, the Congressional Research Service (CRS) characterized disaster assistance pursuant to a presidential declaration as an "entitlement."

> Once the president declares a major disaster, the discretionary nature of the program changes. In essence, it becomes an entitle-

ment program. As long as victims (public or individual) meet eligibility requirements, they are entitled to disaster relief assistance. While this ensures that relief is provided to all victims [regardless of economic need] it may be a potentially expensive arrangement. . . ."[24]

Noting a rising total of declarations per annum, CRS asked whether state and local capacity to handle disasters was in fact diminishing due, presumably, to limits on tax revenue and reliance on federal assistance. It suggested that "Congress may wish to review the criteria used by the president to declare disasters."[25] It also noted the open-ended nature of disaster declarations, which often continue to operate in the designated areas for many years after the disaster. CRS noted that as of September 1991, 184 disaster declarations were "open" and still disbursing funds. As of December 1997, California had 23 disaster declarations still open, dating back to the early 1980s.[26]

Following further catastrophes—the Midwest Flood in 1993 and the Northridge Earthquake in 1994—both the House and Senate formed bipartisan task forces to review federal disaster policies. The declaration process, although not an issue for truly catastrophic events, was thought to be overused in lesser disasters, according to the Senate Task Force. It cited a 1994 audit by the FEMA Inspector General[27] stating that "FEMA has no method for evaluating the capabilities of the state and local governments to effectively respond to the disaster."[28]

The Senate Task Force called for a more objective process than that being used for determining whether or not a presidential declaration should be issued. It also criticized the high probability of obtaining federal assistance for even relatively local disasters: "If the anticipated availability of federal aid induces (more) individuals to locate social or economic activities in hazard-prone areas, then the total annual economic costs are higher with an aid program than without one."[29] The Task Force also cited a 1993 study by the National Academy of Public Administration (NAPA) which asserted that "State and local governments must be able to successfully manage small and medium sized disasters on their own. . . ."[30] The Senate Task Force offered several proposals to reduce federal disaster costs: (1) establish more explicit and/or stringent criteria for providing federal disaster assistance, (2) emphasize hazard mitigation through incentives; and (3) rely more on insurance.[31] (This report is discussed further in Chapter 3.)

Contributing to the distortion of objective consideration of state requests for disaster relief is what NAPA terms the "CNN Syndrome."

One of the most dramatic contextual changes for emergency management is the greater intrusiveness and influence of news media. Disaster and emergencies provide dramatic news and the appetites of news media, particularly television, are insatiable.

This means that emergency management agencies will have to perform under intense media scrutiny. It also means that few emergencies and disaster will remain local—most will now be "nationalized" and politicized as a result of media coverage. . . . The media pressures reluctant local and state leaders to "ask for federal help," presidents to dispatch such help, and representatives and senators to demand it on behalf of constituents.[32]

In 1993, FEMA disaster assistance programs were also reviewed by the National Performance Review (NPR), chaired by Vice President Al Gore. This report echoed others cited earlier in urging that "FEMA's basic role should be to serve as the federal government's coordinator of assistance for state and local governments overwhelmed by disaster, and as a catalyst for development of comprehensive state and local emergency management systems that emphasize loss control and prevention."[33]

Concerning the proliferation of disaster declarations, NPR urged that: "Clear criteria need to be developed for disaster declarations to help conserve federal resources. Federal disaster relief processes should be designed to increase state and local incentives for mitigation (i.e., the use of zoning, building codes, and other damage reduction measures)."[34]

Catastrophes such as Hurricane Andrew, the Northridge Earthquake, and the Midwest Flood account for a large proportion of federal disaster outlays (see Chapter 2). Yet, the overall number of presidential declarations per year has been creeping upward. The average annual total of declarations for 1984–1988 was 26 while the average for 1989–1993 was 42,[35] and for 1993–1997, the average was 45 declarations per year (Table 1-3). President Clinton issued 189 declarations during fiscal years 1996–1998, an average of more than one a week.

The percentage of requests accepted by the White House has generally been rising as well. From the 1950s until 1988, declarations per year averaged about two-thirds of those requested. Since 1988, this proportion has increased to three-quarters, with the Bush and Clinton administrations being the most generous in granting declarations.[36] For fiscal year 1996 (which ended a month before the national election), 72 declarations were issued by President Clinton, comprising an all-time record of 85 percent of those requested (see Table 1-3). The previous record of 83 percent was attained by President Bush in 1992, also a presidential election year. According to University of Delaware political scientist Richard Sylves,[37] the federal government has never applied formal, published criteria to the declaration process. Thus even though very large disasters seem to have increased in frequency and cost, it appears that federal largesse is also broadening to include a variety of smaller disasters that previously might have been considered as within state and local capabilities.

TABLE 1-3 Presidential Major Disaster Declarations
Requested and Granted, Fiscal Years 1984–1997

Fiscal Year	No. Requested	No. Declared	% Declared
1984	48	35	72
1985	32	19	59
1986	38	30	79
1987	32	24	75
1988	25	17	68
1989	43	29	67
1990	43	35	81
1991	52	39	75
1992	56	46	82
1993	51	39	76
1994	51	36	71
1995	45	29	64
1996	85	72	85
1997	66	49	74
1998	n.a.	62	n.a.
Averages:			
1984–1988	35	25	71
1988–1992	44	33	75
1993–1997	60	45	75

Sources: General Accounting Office, 1995 Disaster Assistance: Information
on Expenditures and Proposals to Improve Effectiveness and Reduce Future
Costs. (GAO/T-RCED-95-140), Washington, D.C.: GAO, p. 3; Alan Rhinesmith, Office of Management and Budget. Data provided to National
Research Council Committee on Assessing the Costs of Natural Disasters,
December 15, 1997.
n.a. = not available.

Expanding Federal Costs

Fueled by the potent combination of compassion and politics, congressional disaster outlays rose from the initial $5 million in 1950 to $52 million in fiscal
1953 and have continued to soar after each new catastrophe ever since. The 1995
Senate Task Force estimated that total federal disaster-related expenses, defined
broadly, between 1977 and 1993 amounted to over $119 billion (in constant
1993 dollars).[38] Although not on a par with the defense budget around
$250–300 billion per year, this scale of federal spending on disasters could
scarcely have been foreseen by the drafters of the original act in 1950.

Furthermore, Congress has never funded disaster assistance adequately out of
regular appropriations, which have averaged about $300 million annually dur-

ing the 1990s. Most disaster assistance funding is provided from off-budget "supplementary appropriations" (giving a new twist to the term *supplementary*) according to the catastrophe of the moment. As shown in Table 1-4, some years are relatively economical (e.g., 1983, 1988, and 1991) whereas others are budget-busters (1973, 1978, 1990, 1992, 1993, and 1994). Fiscal years 1988–1997 totaled $30.1 billion in supplemental appropriations for federal disaster assistance. A total of $21.8 billion was allocated by Congress to FEMA for distribution under the Stafford Act. The remaining $8.3 billion provided under supplemental appropriations was allocated to other disaster response agencies, including primarily the U.S. Department of Agriculture, the Department of the Interior, the Small Business Administration, the Army Corps of Engineers, and the Department of Transportation. (See Figure 1-7.) FEMA allocates portions of its Stafford Act funds to procure additional services from these and other mission agencies. The very large allocations to FEMA since FY 1992 reflect the major catastrophes of this decade: Hurricane Andrew, the Midwest Flood, the Northridge Earthquake, and various floods from 1996 to 1998.

The 1995 Senate Task Force report criticized the use of supplemental appropriations to fund disaster assistance on the grounds that (1) assistance is too generous, (2) the practice appears to be fiscally irresponsible, and (3) nonemergency items are funded under disaster funding.[39] The task force recommended:

• Tightening criteria for using the emergency safety valve of the Budget Enforcement Act
• Repealing the automatic exemption of emergency appropriations from budgetary points of order
• Cutting other spending to offset the cost of disaster assistance
• Discouraging funding for nonemergencies in any supplemental containing an emergency designation
• Funding disaster programs at historic average levels
• Creating a Rainy-Day Fund to cover future expenses for federal disaster relief

Until 1996, supplemental appropriations were routinely designated as "emergency" and were therefore exempt from the limits of the 1990 Budget Enforcement Act. In 1996, perhaps in response to the Senate task force report, the 104th Congress offset supplemental appropriations for disaster assistance with rescissions of prior appropriations including $1 billion in unspent FEMA funds.[39] However, the 105th Congress made a further supplemental appropriation of $3.3 billion to FEMA on June 12, 1997, to pay for the costs of the Red River Flood with no rescissions. It is not clear what the future policy of the White House and Congress will be regarding rescissions to cover additional supplemental appropriations.[41] In May 1998, however, rescissions to offset a supplemental appropriation of $2.4 billion were taken from programs not dear to the

TABLE 1-4 Supplemental Appropriations for Disaster Assistance, Fiscal Years 1970–1998 (dollars in millions)

Fiscal Year	Current dollars	Constant 1993 dollars
1970	$305	$1,098
1971	485	1,659
1972	61	198
1973	2,805	8,682
1974	384	1,105
1975	32	345
1976	242	588
1977	904	2,033
1978	3,308	6,924
1979	1,452	2,793
1980	2,797	4,935
1981	233	373
1982	131	195
1983	25	36
1984	153	210
1985	419	553
1986	547	702
1987	109	136
1988	55	66
1989	1,207	1,388
1990	2,850	3,143
1991	0	0
1992	6,063	6,224
1993	3,474	3,474
1994	8,412	8,245
1995	2,275[a]	2,000
1996	3,900[b]	3,600
1997	4,700[c]	4,400
1998	2,400[d]	2,100

Sources: FY 1974–1994: U.S. Senate Bipartisan Task Force on Funding Disaster Relief, 1995, Table 5.1; 1995–1997: Office of Management and Budget; 1998: Legislative History to PL 105-174.

[a] PL 104-19, Ch. X (FEMA only, after $1 billion rescinded by PL 104-134)
[b] PL 104-19 (FEMA share: $3.1 billion)
[c] PL 105-18, Ch. IX (FEMA share: $3.3 billion)
[d] PL 105-174 Appropriations for disaster relief and Bosnia peacekeeping forces offset by rescissions from low-income housing, Americorps, bilingual education, and airport funds

Republican leadership, namely low-income housing, Americorps, and bilingual education (Table 1-4).

Supplemental appropriation totals, however, do not tell the whole story of federal assistance. They do not cover on-budget disaster appropriations, subsidized loans and insurance payments, and other federal expenditures relating to disasters not pursuant to a disaster declaration.

The magnitude of federal liability for disaster assistance obviously depends in part on the share of such costs assumed by nonfederal interests, chiefly state and local governments. No cost-sharing formula was established until the 1974 Disaster Relief Act prescribed a 25 percent nonfederal share for individual and family grants.[42] The nonfederal contribution to all other assistance remained subject to negotiation until May 1980, when FEMA administratively adopted a general policy of requiring state and local governments to agree to pay 25 percent of the eligible costs of public assistance programs.[43] This policy removed the administrative problem associated with attempting to determine a "reasonable" commitment for each disaster. Although hotly challenged by members of Congress, this established the standard ratio of cost-sharing that was later incorporated into the 1988 Stafford Act. In effect, however, it belies the "supplementary" federal role by codifying that the bulk of disaster costs will be federal. Also the nonfederal share often is covered by "in kind" or "soft" match, instead of new allocations.[44] Sometimes even federal money can be used for the nonfederal share. After the Midwest Flood of 1993, federal community development block grant (CDBG) funds were used for this purpose. (See Chapter 7.)

Even the 25 percent nonfederal share has proven negotiable. In approximately 15 severe disasters since 1985, the requirement of a 25 percent nonfederal share for matching assistance to state and local governments has been waived to varying degrees. The nonfederal share for public assistance costs after the Northridge Earthquake was set at 10 percent and for Hurricane Andrew, at zero.[45] (A nonfederal cost share for individual assistance, however, cannot be waived.)

THE BROADER UNIVERSE OF FEDERAL DISASTER PROGRAMS

Federal assistance provided from the president's Disaster Relief Fund under the Stafford Act accounted for 80 percent of the $5.4 billion in obligations by FEMA in fiscal year 1994.[46] But it comprises only an indeterminate fraction of total federal disaster-related assistance provided in that and other fiscal years. Approximately 30 federal programs of many types offer some form of disaster service or funding.[47] Twenty-six federal departments and agencies with disaster response functions, plus the American Red Cross, are encompassed by the Federal Response Plan coordinated by FEMA.[48]

There is no systematic compilation of the total federal costs relating to disasters. In part this is a definitional problem since agency budgets may not clearly define which expenditures are disaster-related and which are not. It also involves the problem of adding up apples and oranges: While the bulk of federal costs are in the form of grants, other commitments are made as low-interest loans and as payments from government-sponsored insurance programs whose costs are defrayed to varying degrees by premiums paid. There is also a political explanation for the lack of a comprehensive federal database on disaster costs: Congress and the president may prefer to keep the true extent of federal outlays obscure to avoid charges of "pork barrel politics" arising from the uneven and sometimes politically motivated allocation of disaster declarations and program outlays. (See Chapter 2.)

Perhaps the only recent attempt to calculate the total universe of federal spending on disasters was the 1995 report of the Senate Bipartisan Task Force on Funding Disaster Relief.[49] The task force estimated federal disaster-related expenditures from 1977 through 1993 devoted to the four stages of disaster assistance: (1) preparedness, (2) emergency response, (3) recovery, and (4) mitigation (reduction of future vulnerability). For the 16-year period under consideration, the task force estimated the grant total of federal spending in all four phases amounted to approximately $119 billion in 1993 dollars (Table 1-5). Of that amount, direct grants to communities and individual victims, including but not limited to the Disaster Assistance Program, amounted to about $64 billion. The remainder of the $119 billion, approximately $55 billion, took the form of low-interest loans and insurance payments. In this case, the actual tax cost is much less than the total outlays, considering repayments of loans and payment of insurance premiums. The actual federal cost to taxpayers is represented by the amount to which insurance premiums are subsidized and loans are issued at subsidized rates or are forgiven (i.e., not required to be repaid).

Since 1950, Congress has established several disaster-related programs outside the framework of the basic disaster assistance acts. These programs are not tied to a presidential declaration and several involve loans or insurance to relieve the burdens on the federal treasury of outright grants. These include principally the National Flood Insurance Program (NFIP), the Small Business Administration (SBA) disaster loan program for homes and businesses, the Farmers Home Administration (FmHA) disaster loan program, the Federal Crop Insurance Pro-

TABLE 1-5 Federal Expenditures for Disaster Assistance: Grants, Loans, and Insurance for Fiscal Years 1977–1993 (578 declarations)

Type of Assistance	1993 Dollars (millions)	Percent of Expenditure (%)
Preparedness	$2,232	2
Mitigation	$27,073	22
Response	$3,377	3
Recovery	$86,989	73
TOTAL	$119,671	100

Source: U.S. Senate Bipartisan Task Force, 1995, Table 1.1

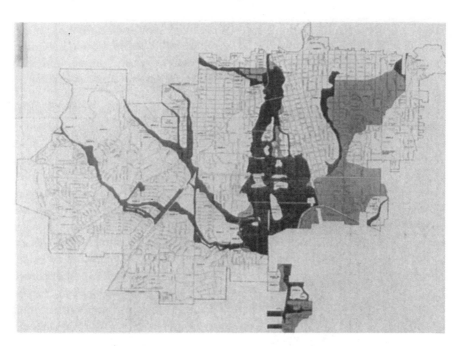

FIGURE 1-4 A FIRM for Boulder, Colorado, ca. 1980 showing 100-year (dark shading) and 500-year (light shading) flood hazard areas.

Coast and the flash flood in Rapid City, South Dakota—Congress passed the Flood Disaster Protection Act of 1973.[57] This act enlarged the scope of the incipient NFIP and required purchase of a flood insurance policy by anyone receiving federally related financing involving floodprone property. It also added mudslides and erosion as insurable hazards under the NFIP.

The National Flood Insurance Program has expanded rapidly since 1973. As of September 30, 1998, over 4.1 million policies covering more than $482 billion in floodprone structures and their contents were in force. Coverage increased by 14.3 percent during fiscal year 1997 as a result of aggressive marketing of flood insurance by FEMA and the prospect of heavy flooding on the West Coast due to El Niño. During seven years between 1984 and 1997, the NFIP took in more revenue than it paid out. But in the other seven years, including all but one since 1992, the program's outlays exceeded its revenue, resulting in net borrowing from the U.S. Treasury. As of the end of FY 1997, the program was in the red for a total of $1.1 billion (Table 1-6).

Yet only an estimated 20 to 25 percent of floodprone properties are covered under the program nationally. In the Midwest Flood of 1993, only about one out of five flooded structures was insured. Total NFIP payments in that disaster

TABLE 1-6 National Flood Insurance Program: Policies, Coverage, and Net Income or Loss: FY 1974–1997

Fiscal Year	No. of Policies (1000s)	Coverage ($ billions)	Net Income or (Loss)
1974	385	$8.4	n.a.
1979	1650	$62.5	n.a.
1984	1831	$115	($61 mill.)
1985	1955	$133	$273 mill.
1986	2075	$133	$29.7 mill.
1987	2079	$158	$171 mill.
1988	2101	$169	$183 mill
1989	2200	$179	($146 mill.)
1990	2378	$203	$73 mill.
1991	2506	$219	$169 mill.
1992	2561	$229	($20.4 mill.)
1993	2725	$254	(602 mill.)
1994	2804	$274	$269 mill.
1995	3264	$325	($576 mill.)
1996	3546	$369	($536 mill.)
1997	3811	$422	($117 mill.)
1986–1997 (CUM. TOTAL)			($1.1 billion)

Source: Federal Emergency Management Agency/Federal Insurance Administration data

amounted to $293 million, the third highest payout in the history of the program (behind Hurricane Hugo in 1989 and the Northeast winter storms of 1992–1993). But this still amounted to only 2 percent of estimated total losses of $12 billion in the Midwest Flood, and only 7 percent of direct federal costs paid by taxpayers amounting to $4.2 billion.[58] (See Chapter 7.)

The NFIP is heavily coastal. (See Tables 1-7 and 1-10.) About 58.7 percent of policies in force and 62.8 percent of insurance coverage in force pertain to coastal communities. As defined by the NFIP, these include municipalities and counties (in the case of unincorporated areas) that border on coastal waters. An unknown portion of the coverage in these communities applies to inland flood hazards, but presumably the largest share of the policies and value of coverage is on or near the actual coastlines. The program is greatly in demand by the owners of recent and opulent development at the ocean's edge (Figures 1-5 and 1-6).

One of the most serious weaknesses of the National Flood Insurance Program is the problem of repetitive losses. The program has been notoriously willing to continue insuring structures despite repetitive claims, often with little increase in premium. According to a recent study by the National Wildlife Federation, 74,500 properties (2 percent of insured properties) have accounted for 200,182

TABLE 1-7 National Flood Insurance Program Activity: Coastal and
Noncoastal Communities (as of March 31, 1997)

	Coastal Communities (% of NFIP Total)		Noncoastal Communities (% of NFIP Total)	Total NFIP
	Total Community[b]	V-Zone Only		
Policies in Force	2.1 mill. (58.7%)	77,298 (2.1%)	1.5 mill. (41.3%)	3.6 mill.
Insurance in Force	$247 bill. (62.8%)	$4.1 bill. (2.2%)	$146 bill. (37.2%)	$393 bill.
Total Premium	$758 mill. (63.1%)	$27.5 mill.	$522 mill. (36.9%)	$1.2 bill.
Average Premium	$352	$720	$343	$348
Average Coverage	$114,520	$115,615 (post '81) $103,327 (other V)	$96,230	$106,960
No. of Claims[a]	316,472 (48.8%)	24,084 (3.7%)	331,942 (51.2%)	648,414
Amount of Paid Losses[a]	$4.1 bill. (54.8%)	$406 mill. (5.4%)	$3.4 bill. (45.2%)	$7.5 bill.
Average Loss[a]	$12,972	$25,359 (post '81) $14,645 (other V)	$10,215	$11,561

Source: Federal Emergency Management Agency (unpublished data)
[a] January 1, 1978 through March 31, 1997
[b] Defined by FEMA as all communities containing a "coastal high hazard area" (V-zone); includes inland portions of such communities

claims, comprising over one-quarter of all NFIP claims. *Payments for repetitive loss properties have amounted to $2.58 billion, or 40 percent of all NFIP payments.* Approximately 37 percent of repetitive loss properties flooded three or more times with payments totaling $1.4 billion (53 percent of repetitive loss payments).[59]

The repetitive loss problem reveals a major difference between governmental and private insurance plans. While the latter would either cancel coverage or raise premiums significantly after repetitive losses, the government is reluctant to offend anyone. Indeed, there is substantial lobbying by private property interests to protect availability of NFIP coverage at reasonable rates, regardless of actual loss experience. Federal flood insurance thus is equivalent to an "entitlement" that property owners claim as a right, even when it is subsidized by taxpayers. (See Chapter 6.)

FIGURE 1-5 A nearly beachless Ocean City, Maryland, in the mid-1980s. This resort area within a day trip of Washington, D.C., and Baltimore has since undergone repetitive beach nourishment projects. (Photo by R.H. Platt)

FIGURE 1-6 New construction at Sullivan's Island, South Carolina, ca. 1991, on a shore recently ravaged by Hurricane Hugo. These structures are elevated, as required by the NFIP, and are eligible for flood insurance although perched precariously on an eroding beachfront. (Photo by R.H. Platt)

Small Business Administration Disaster Loans

The Small Business Administration (SBA) Disaster Loan Program was initiated by Congress in 1953. Unlike other SBA activities, disaster loans are authorized to be issued not only to small businesses but also to homeowners, tenants, and nonprofit organizations. Loans are made to such entities that are victims of disasters (as declared by either the president or by the administrator of SBA) to help cover uninsured losses. Most disaster loans are made at below-market interest rates (currently about 4 percent) for long terms (up to 30 years), thus reflecting a federal subsidy as compared with market loans. SBA expects to be repaid and refers applicants who lack the ability to repay a loan to FEMA for a possible "Individual and Family Grant," available to low-income households after a presidentially declared disaster. SBA secures its loans, when possible, with liens on the relevant real property and may foreclose in the event of nonpayment. However, its claim would be subordinate to those of prior mortgage lenders.[60]

Since 1953, SBA has approved slightly over 1.4 million disaster loans for a total of $25.5 billion. Loans to homeowners and tenants account for 77 percent of the number of loans and 41 percent of the dollar value, with the remainder largely extended to small businesses. The largest disaster event in the history of the program, not surprisingly, was the Northridge Earthquake of 1994, which contributed to a total for that fiscal year of 125,861 loans amounting to $4.1 billion, about four times the recent annual average of about one billion dollars.[61] The actual loss rate on SBA disaster loans over the life of the program is 8.27 percent. Table 1-8 shows the number of disaster loans made by SBA since 1988.

TABLE 1-8 Small Business Administration Disaster Loans

Fiscal Year	No. of Loans	Value of Loans ($millions)
1988	13,456	$234
1989	5,330	148
1990	51,970	1,320
1991	12,451	356
1992	23,417	794
1993	58,644	1,673
1994	125,861	4,159
1995	45,041	1,217
1996	37,822	987
1997	49,515	1,138
1998*	20,642	495
Total (1988–98*)	444,149	$12,521
Total (1953–98*)	1,407,042	$25,587

Source: Small Business Administration (unpublished data)

*Through May 31, 1998

U.S. Department of Agriculture Disaster Programs

Agriculture is an inherently risky occupation, in which the federal government has long played a paternalistic role. Under a myriad of different authorities, the U.S. Department of Agriculture (USDA) provides crop insurance, farm disaster payments, and emergency disaster loans to farm-related businesses. In the midst of depression and drought, the Federal Crop Insurance Corporation was created by the Agricultural Adjustment Act of 1938 (PL 75-430) to insure wheat producers against unavoidable losses.[62] Until the Federal Crop Insurance Act of 1980 (PL 96-365), the federal crop insurance program applied only to selected crops in particular counties, with no subsidy in the premium charged to farmers. It was thus a little-used tool for protection against disaster-related crop losses. The 1980 act sought to expand coverage and to subsidize premiums. Farmers still resisted signing up for crop insurance, in part because outright disaster payments were becoming increasingly common. By the time of the Midwest Flood of 1993, participation in the crop insurance program was about 35 percent nationally.[63] In any event, the crop insurance program was not a self-supporting program. Between 1977 and 1993, the program paid insured farmers a total of $13.4 billion at a net cost to the federal taxpayer of $9.9 billion. During the same period, the Farmers Home Administration (FmHA, now the Farm Service Agency) extended $134 million in disaster loans.[64]

These programs still failed to serve the demand in Congress to indemnify farmers against disaster losses. Beginning in 1988, Congress has approved a series of ad hoc supplemental appropriations to fund agricultural disaster payments among other federal disaster benefits programs. Payments are available to farmers who experience a greater than 40 percent loss in production in a given year. The amount of the payment is the difference between that 40 percent "self-absorbed" loss and the market price of the commodity in question. This has resulted in some very large outlays of federal dollars to farmers as direct grants, as opposed to insurance or loans. Under the 1988 act, USDA made cash disaster payments of $3.37 billion to farmers. Several billions more have been expended under subsequent appropriations. After the Midwest Flood of 1993, the USDA disbursed approximately $1.4 billion in disaster payments, in addition to $1.0 billion in crop insurance payments and $14.7 million in emergency disaster loans through the FmHA.[65]

The prospect of outright disaster payments discouraged farmers from signing up for the annual costs of participation in the crop insurance program. Congress was undermining its own efforts to move disaster costs toward a self-funding basis by its liberal funding of disaster grants with no contribution by the recipient. However, a policy review after the Midwest Flood of 1993 led to adoption of PL 103-354 in 1994, which provides the following:

• Mandatory participation if farmers are to receive any other USDA subsidies
• A catastrophic coverage policy for farmers

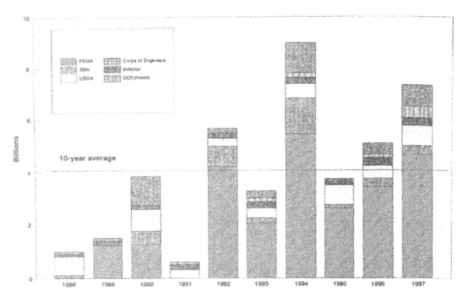

FIGURE 1-7 Major disaster assistance authorizations: fiscal years 1988–1997 (in billions, not adjusted for inflation). Compiled by the Office of Management and Budget for the National Research Council Committee on Assessing the Costs of Natural Disasters, December 1997.

As early as 1973, a White House policy paper recognized that federal disaster assistance had become counterproductive by *replacing* rather than *supplementing* nonfederal efforts.

> The series of disaster assistance laws that culminated in the passage of PL 93-385 in August, 1972 made the federal government virtually the sole provider of recovery assistance in case of a major disaster. The concept of such assistance being supplementary to that of the state and local governments remained as an expression of intent, but was less operative in practice. *Also the benefits were so generous that individuals, businesses, and communities had little incentive to take initiatives to reduce personal and local hazards*[71] (emphasis added).

These benefits have often been provided without significant strings attached to reduce vulnerability to future losses to the same property or community. Repetitive losses to the same insured properties account for more than 40 percent of NFIP payments, according to a recent study by the National Wildlife Federation.[72]

In 1985, Bruce B. Clary characterized federal disaster policy as "a textbook example of incremental decision making." With reference to Lowi's typology of public policies as distributive, redistributive, and regulatory, Clary viewed federal disaster policies before the late 1960s as "almost entirely distributive" and counterproductive.

> The pattern of increasing benefits, along with progressively more liberal eligibility criteria and decreasing interest rates for federal loans, continued into the mid-1970s. By this time the federal share of disaster costs had risen to 70 percent (from 1 percent in 1953). The net effect was to lessen significantly the likelihood of large property loss from a natural disaster. But, in many cases, it promoted poor locational decisions, since the federal government was subsidizing much of the risk.[73]

Beginning in the late 1960s, particularly with the creation of the National Flood Insurance Program in 1968, Clary identified a shift in federal policy toward "an increasing shift toward regulation." He viewed this as a desirable trend, especially as accompanied by the imposition of a minimum 25 percent nonfederal cost share for disaster assistance.

Yet, in 1994, the Executive Summary of the House Bipartisan Natural Disasters Task Force[74] echoed the 1973 White House report: "If state and local governments believe that the federal government will meet their needs in every disaster, they have less incentive to spend scarce state and local resources on disaster preparedness, mitigation, response, and recovery. This not only raises the costs of disasters to federal taxpayers, but also to our society as a whole as people are encouraged to take risks they think they will not have to pay for."[75]

Repeated flooding in California in the winter of 1995, which resulted in the declaration of all counties in that state as disaster areas, reinforced pessimism regarding the effectiveness of past efforts to reduce vulnerability. During the first round of these floods, a front page New York Times article reported criticism of federal policies under the headline: "California Storm Brings Rethinking of Development—A Cycle of Destruction—Some Officials Begin Exploring How to Keep People From Building in Flood Zones." The article reported that:

> Critics of federal policy on disaster relief say it has long encouraged people to stay in areas where homes probably would not be built. . . . There is no incentive for a homeowner to buy insurance or to move out of harm's way, critics say, when flood

victims know they can rely on federal grants and loans after a flood.[76]

Although this newspaper article was based largely on the views of environmentalists, it was reinforced by testimony in February 1995 on behalf of the Association of State Floodplain Managers.

> If we as a nation are going to bring escalating disaster costs under control we need to embrace the concept of hazard mitigation. Our flood policies have not embraced hazard mitigation and in part are to blame for escalating disaster damages. On an individual project basis, flood control projects have reduced flooding for design floods. But at the same time these policies have enticed additional development increasing the damage potential for severe floods, or have silently promoted the transfer of flood damages from one property to another. Likewise, with a benevolent federal government, there has been little incentive at the local and state levels of government to minimize the creation of new hazards.[77]

Such views regarding the counterproductive impact of federal disaster programs has stimulated reconsideration of the role of hazard mitigation as reflected in the creation of a FEMA Mitigation Directorate in 1994 and the drafting of a National Mitigation Strategy (FEMA 1995).[78] The evolution and debates concerning the meaning of "mitigation" will be examined in Chapter 3.

Most commentators on federal disaster relief policy seem to agree implicitly that it is appropriate for the federal government to provide benefits to disaster victims, subject to appropriate conditions. Seldom is the wisdom of federal disaster relief per se called into question. In a departure from the prevailing view, philosopher William B. Irvine asks why victims of natural disasters have a "moral claim" on federal tax money, whereas victims of other personal losses, such as a house fire or a stock market loss, have no such claim. Irvine argues that reliance for recovery should be placed with private insurance coverage (which rebuilt downtown San Francisco after the 1906 Earthquake) and with charitable philanthropy. He would thus return to the period before 1950 when disaster recovery was not a federal function.

> When the government adopts (either explicitly or implicitly) a policy of making such compensation [for natural disaster losses], it inadvertently sets the stage for even greater disasters in the future. . . . If, after all, the government adopts a policy of bailing out those who lose the bet they place when they pass up earthquake insurance or build their home on a beach, the government

unintentionally encourages people to engage in this sort of behavior ("why pay for disaster insurance when you can get it for free from the government?"). In other words, disaster relief today can result in additional disasters tomorrow.[79]

This perception is reflected in the following comment of a real estate developer at North Topsail Beach, the scene of major damage due to Hurricanes Bertha and Fran in 1996, and a reprise by Hurricane Bonnie in 1998: "Do you realize the risk that people take every day when people invest in the stock market? People every day take much more risk than buying property on a beautiful beach."[80]

In a similar vein, but from a homeowner's point of view, ABC News reporter John Stossel, the recipient of federal flood insurance payments for the loss of his oceanfront home, considers such assistance to be "welfare for the wealthy."

> In 1980 I built a beach house on the edge of the ocean on Long Island. It was an absurd place to build a house. One block down the road was a desert-like wasteland where a dozen houses had eroded away a few years earlier. Still, my eager-for-the-business architect said, Why not build? If the ocean destroys your house, the government will pay for a new one.
>
> It seemed odd that the government would encourage people to take foolish risks, but the contractor was right. If the ocean took my house, Uncle Sam would happily pay to replace it (up to $350,000). Thanks, taxpayers! It's called national flood insurance. . . . During a fairly ordinary storm, the ocean knocked down my government-approved pilings and ate my house. It was an upsetting loss for me, but financially I made out just fine. Federal flood insurance paid for the house—and its contents.[81]

Finally, as an ironic commentary on how far the pendulum of federal disaster response has swung since 1950, the following editorial entitled "Awash in Tax Dollars" appeared in the Raleigh, North Carolina, *News & Observer* actually *objecting* to the overgenerosity of federal assistance to coastal communities struck by Hurricane Fran a year earlier.

> In the Wake of Hurricane Fran, North Carolina's coastal communities and residents mined taxpayer accounts to rebuild in fragile areas. Such generosity encourages overdevelopment, at great expense. . . . The allocation of hundreds of millions in taxpayer dollars has led the federal government to undermine what state officials have been trying to do for decades—discourage development in coastal areas that are vulnerable not just to hurricanes but to heavy storms of any kind.[82]

28. U.S. Senate Bipartisan Task Force on Funding Disaster Relief (1995). *Federal Disaster Assistance.* Washington, D.C.: U.S. Government Printing Office, 70.

29. Ibid.

30. Ibid., 53, citing National Academy of Public Administration (NAPA) (1993). *Coping with Catastrophe.* Washington, D.C.: NAPA, xi.

31. General Accounting Office (1995). Statement of Judy A. England-Joseph to Senate Committee on Appropriations, Subcommittee on VA, HUD, and Independent Agencies on "Information on Expenditures and Proposals to Improve Effectiveness and Reduce Future Costs" (GAO/T-RCED-95-140). Washington, D.C.: GAO.

32. NAPA (1993), note 30, 18.

33. National Performance Review (1993). *Creating a Government that Works Better and Costs Less: FEMA.* Washington, D.C.: Office of the Vice President, 1.

34. Ibid.

35. GAO (1995), note 31.

36. Richard T. Sylves, Presentation to National Research Council Committee on Assessing the Costs of Natural Disasters, December 15, 1997.

37. Sylves, note 11, p. 34.

38. Senate Task Force, note 28, Table 1.1.

39. Ibid., 84.

40. Omnibus Consolidated Rescissions and Appropriations Act of 1996 (PL 104-134).

41. Alan Rhinesmith, Office of Management and Budget. Presentation to National Research Council Committee on Assessing the Costs of Natural Disasters, December 15, 1997.

42. PL 93-288, sec. 408(b).

43. GAO (1981), note 16, 36.

44. Ibid.

45. Senate Task Force, note 28, 15.

46. GAO (1995), note 31, 2.

47. Bea, note 24, 11.

48. Federal Emergency Management Agency (1992). *Federal Response Plan.* Washington, D.C.: U.S. Government Printing Office.

49. Senate Task Force, note 28.

50. Federal Emergency Management Agency (1981). *Evaluation of the Economic, Social and Environmental Effects of Floodplain Regulations.* Washington, D.C.: FEMA, 1.

51. U.S. Congress (1966). *Insurance and Other Programs for Financial Assistance to Flood Victims.* Senate Committee on Banking and Currency (89th Cong., 2d sess.) Washington, D.C.: U.S. Government Printing Office.

52. Task Force on Federal Flood Control Policy (1966). *A Unified National Program for Managing Flood Losses.* House Doc. 465 (89th Cong. 2nd sess.) Washington, D.C.: U.S. Government Printing Office.

53. Rutherford H. Platt (1986). "Floods and Man: A Geographer's Agenda," in Robert W. Kates and Ian Burton, eds., *Geography, Resources, and Environment* (Vol. 2). Chicago: University of Chicago Press, 28–68.

54. PL 90-448, Title XIII. The National Flood Insurance Act is codified at 42 USCA 4001 et seq.

55. May, note 5, 19.

56. Rutherford H. Platt (1976). "The National Flood Insurance Program: Some Midstream Perspectives." *Journal of the American Institute of Planners* 42(3): 303–313.

57. PL 93-234.

58. Rutherford H. Platt (1995). "Report on Reports: Sharing the Challenge: Floodplain Management into the 21st Century." *Environment* 37(1): 25–29, at 26.

59. David Conrad (1998). *Higher Ground.* Washington, D.C.: National Wildlife Federation.

60. Senate Task Force, note 28, 102–108; Becky Brantley, SBA Loan Officer, pers. comm. June 19, 1998.

61. Small Business Administration, unpublished data, as of May 31, 1998.

62. Senate Task Force, note 28, 119.

63. Thomas P. Zacharias (1996). "Impacts on Agricultural Production: Huge Financial Losses Lead to New Policies," in Stanley A. Changnon, ed., *The Great Flood of 1993: Causes, Impacts, and Responses.* Boulder, Colo.: Westview Press, 163–182, at 181.

64. Senate Task Force, note 28, Tables 1.7 and 1.9.

65. Interagency Floodplain Management Review Committee (1994). *Sharing the Challenge: Floodplain Management into the 21st Century.* Washington, D.C.: U.S. Government Printing Office, Tables 1.5, 1.6, and 1.9.

66. Zacharias, note 62, 181.

67. Ibid., 182.

68. PL 93-303, Title 1, sec. 101(c).

69. U.S. General Accounting Office (1996). *Disaster Assistance: Improvements Needed in Determining Eligibility for Public Assistance,* (GAO/RCED-96-113). Washington, D.C.: GAO, 15.

70. Kenneth Deutsch, American Red Cross. Pers. Comm., July 7, 1998.

71. President of the United States (1973). *New Approaches to Federal Disaster Preparedness and Assistance.* House Doc. 93-100 (93rd Cong., 1st sess.). Washington, D.C.: U.S. Government Printing Office, 5.

72. Conrad, note 58.

73. Bruce B. Clary (1985). "The Evolution and Structure of Natural Hazard Policies," *Public Administration Review* (45): 20–28.

74. House of Representatives Bipartisan Natural Disasters Task Force (1994). *Report.* (Dec. 14). Mimeo.

75. Ibid., 1.

76. *The New York Times* (1995). "California Storm Brings Rethinking of Development" (January 15): 1 and 20.

77. Doug Plasencia, Testimony of the Association of State Floodplain Managers before the Senate Subcommittee on Transportation and Infrastructure, February 14, 1995 (mimeo).

78. Federal Emergency Management Agency (1995). *National Mitigation Strategy: Mitigation . . . Reducing the Impact of Natural Hazards.* Washington, D.C.: FEMA.

79. William B. Irvine (1990). "Saying No to Federal Disaster Relief." *Freeman* 40(3): 104–105.

80. John Starling, real estate developer, North Topsail Beach, N.C., as quoted in *The News & Observer* (Raleigh, N.C.), December 21, 1989.

81. John Stossel (1995). "Swept Away by Welfare for the Wealthy." *The Wall Street Journal* (August 10): 14.

82. Editorial: "Awash in Tax Dollars." *News & Observer,* November 11, 1997, 14A.

CHAPTER 2

U.S. Federal Disaster Declarations: A Geographical Analysis

Ute J. Dymon[1] and Rutherford H. Platt

INTRODUCTION

Average annual loss of life from natural disasters in the United States fortunately has decreased during the twentieth century, in part due to improved warning and emergency response capabilities.[2] However, the rising economic costs of disasters to government at all levels and to society have stimulated widespread concern in Congress and among informed observers.[3] As discussed in Chapter 1, the rapid rise in economic costs of natural disasters have been fueled in part by an increase in the number of declared disasters, the occurrence of a series of unusually large disasters, and the broadening of allowable categories of federal assistance. FEMA outlays for public assistance, the largest category of funding provided under the Stafford Act, totaled $1.0 billion during fiscal years 1983–1988 but amounted to $6.5 billion for fiscal years 1989–1994 (both time periods in constant 1995 dollars).[4] Similarly, the average annual number of declarations issued rose from 26 for the period 1984–1988 to 45 for 1993–1997 (Table 1-3).

The upward trend of disaster costs, bemoaned on all sides, is by no means clear-cut, however. As Figure 2-1 indicates, the level of Stafford Act outlays displayed two peaks between 1989 and 1995 and otherwise was fairly confused. Those peak calendar years of 1992 and 1994 reflected Hurricane Andrew and the Northridge Earthquake, respectively. The question naturally arises as to whether those catastrophes were outliers or exceptions to "normal" disaster experience, or are they tocsins of future catastrophes. In the midst of the bizarre weather attributed El Niño in 1998, one is tempted to choose the latter. Indeed, a future Southern California or Bay Area Earthquake, expected in the next 30 years or so, could make Northridge look insignificant. The January 17, 1995, earthquake in Kobe, Japan, which is widely regarded as a model for the next major California earthquake, inflicted 6308 deaths and 43,177 injuries, and damage exceeding $100 billion.[5]

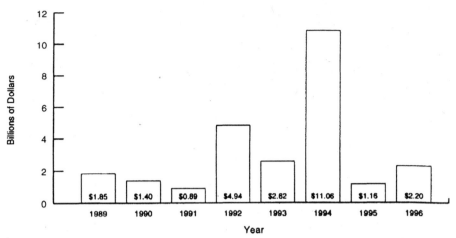

FIGURE 2-1 Total Stafford Act outlays per year: January 1989 to December 1995, in constant 1995 dollars. (Amounts are shown according to year when obligated.) (Data source: Federal Emergency Management Agency. Graphics by Ute Dymon.)

The actual magnitude of U.S. federal disaster costs is unknown. There is no central database that records government-wide disaster outlays. FEMA accounts for its own outlays under the Stafford Act and the National Flood Insurance Program. Stafford Act funds include FEMA's own direct outlays as well as transfers of funds to other agencies under mission assignments. However, many federal agencies such as the Departments of Agriculture and Transportation, and the Army Corps of Engineers respond to disasters partly under Stafford Act transfers from FEMA and partly with their own resources. Disaster- and nondisaster-related costs of such agencies are difficult to unravel. Furthermore, federal outlays take many forms that defy summation: grants, reimbursements, subsidies, loans, insurance—the "apples and oranges" problem. Also, federal cost data are usually reported by fiscal year and often are not totaled for particular disasters for which federal response costs extend over multiple fiscal years. The principal effort to date to compile a government-wide estimate of federal disaster costs was conducted by the Senate Bipartisan Task Force in 1995.[6] As mentioned in Chapter 1, that study estimated that the total cost of federal disaster-related programs between 1977 and 1993 amounted to over $119 billion (in 1993 dollars).

Occasionally, the federal government takes a closer look at its actual costs relating to a specific disaster. For example, a special interagency committee was appointed by the White House after the Midwest Flood of 1993 to assess its impacts and to recommend ways to reduce future vulnerability. The ensuing report identified $1.2 billion expended under the Stafford Act. However, the actual federal cost of that disaster (involving declarations for nine Midwest states) was estimated to be approximately $4.2 billion "in direct federal expenses" plus $1.3 billion in payments from federal insurance programs (crop and flood insurance) and more than $621 million in federal loans to individuals, businesses, and communities.[7] (See Chapter 7.) Presumably the additional "direct federal expenses" involved various agency commitments outside the umbrella of the Stafford Act, and possibly not even known to those who maintain FEMA's database. Such widely varying estimates are pervasive in disaster cost accounting: It is virtually impossible to total up all the diverse federal costs, direct and indirect, that pertain to each disaster declaration.

Even more unfathomable is the scope of the total costs that disasters inflict on society, including private insurance payments, noninsured losses, indirect costs, and nonmonetary impacts.[8] Private insurance industry payouts for Hurricane Andrew exceeded $16 billion, causing nine insurance companies to become insolvent. Two years later, the insurance industry paid $12.5 billion due to the Northridge Earthquake.[9] But, like the federal government, the insurance industry lacks an ongoing, comprehensive process for tracking disaster costs. Corporate data on claims paid is frequently treated as proprietary and is not disclosed.

Many disaster costs are indirect or secondary, such as losses to businesses that are closed or are inaccessible to customers. Disruption of transportation and other lifelines inflicts incalculable costs on the economy, on institutions such as educational and medical facilities, and on social well-being. Disasters also yield nonquantifiable costs to natural resources, the environment, and to scenic amenities, such as the loss or decapitation of millions of trees in the 1998 Northeast ice storms. Many costs are simply borne by the victims and are unrecorded and uncounted in any estimate of disaster costs—the loss of treasured family photographs, a pet, or a graduate thesis in progress.

Disaster costs are sometimes overestimated by zealous politicians in the midst of a disaster. At best, these are likely to be "guesstimates" based on a helicopter overflight and back-of-envelope calculations or a quick GIS computer run. At worst, they are gross exaggerations calculated to attract a presidential disaster declaration. In either case, the more horrific the number, the more likely it will be trumpeted by the media and become accepted as "fact." Public policy is disserved as much by overestimates as by underestimates of the actual costs of disasters. (FEMA and state officials are charged with preparing preliminary damage assessments (PDAs) as a more objective basis for requesting a presidential declaration.)

Our knowledge of the total extent and types of disaster costs in the United States (and, perforce, the world) is thus fragmentary and incomplete. But viewing such costs, no matter how calculated, from a national perspective overlooks a further important qualification, namely the spatial diversity of disasters and their impacts. Disasters are not ubiquitous, and neither is the incidence of federal costs, insurance payments, social burdens, and environmental harms that they generate. Some places or regions are more hazardous than others and experience much greater losses. Furthermore, the spatial distribution of federal disaster assistance may vary according to the political significance of a particular state or the influence of a particular member of Congress. There is thus a physical/political geography of the incidence of disaster costs that varies over time.

Perhaps the best barometer of the fluctuations of disaster costs over time and space within the United States is the record of costs incurred by FEMA under the 1988 Disaster Relief and Emergency Assistance Act (Stafford Act). While an imperfect measure of total costs, for reasons stated above, Stafford Act outlays are the "mainstream" of federal disaster assistance. For present purposes, they are assumed to reflect the pulses of overall costs attributable to disasters. Furthermore, Stafford Act outlays are susceptible to spatial analysis, since disaster declarations pertain to specific states and discrete counties within states.

Stafford Act activity is strongly dominated by a few catastrophic events that produce dramatic fluctuations from year to year (Figures 2-1 and 2-2). About three-fourths of all federal disaster assistance under the Stafford Act has gone to four states (California, Florida, Georgia, and South Carolina) plus Puerto Rico. All other states and territories have collectively received about one-fourth of Stafford Act outlays. This is about the same proportion as the dollars allotted to all disaster declarations other than the top five as shown in Figure 2-2.

But Stafford Act assistance also extends to an ever-wider range of smaller disasters in step with the growing enthusiasm of presidents for issuing declarations. It is a reasonable indicator of the magnitude and location of most of the nation's disaster costs.

The following analysis therefore employs FEMA Stafford Act data as a surrogate for the broader range of federal and nonfederal disaster costs. Computer-drawn maps prepared by the Kent State University GIS Lab reveal spatial patterns of declarations since 1989, the allocation of federal funds by state and county, and by type of disaster. The graphics provide a broad overview of the spatial distribution of Stafford Act activity, but they do not include other federal programs not tied to the Stafford Act, such as National Flood Insurance payments, Small Business Administration disaster loans, Department of Housing and Urban Development block grants, or Department of Agriculture disaster payments to farmers.

**Largest Total Outlays
by State/Territory**

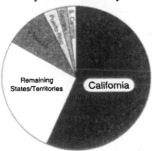

Remaining
States/Territories

California

Total Payments = 21.61 Billion

California: $12.251 (56.69%)
Florida: $2.009 (9.30%)
Puerto Rico: $0..679 (3.14%)
Georgia: $0.456 (2.11%)
South Carolina: $0.404 (1.87%)
Remaining States: $5.811 (26.89%)

**Five Most Costly
Disaster Declarations**

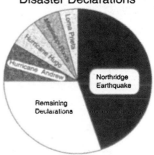

Northridge
Earthquake

Remaining
Declarations

Payments in billions of dollars

263 Declarations

Northridge Earthquake: $9.713 (44.97%)
Hurricane Andrew: $1.9 (8.79%)
Hurricane Hugo: $1.38 (6.39%)
Mississippi River Floods: $1.253 (5.80%)
Loma Prieta Earthquake: $0.853 (3.95%)
Remaining Declarations: $6.506 (30.10%)

FIGURE 2-2 Largest Stafford Act outlays by state and declaration: January 1989 to December 1995, in billions of dollars. (Data source: Federal Emergency Management Agency. Graphics by Ute Dymon.)

SPATIAL PATTERNS OF STAFFORD ACT ASSISTANCE

The State Level

Figure 2-3 represents the state-by-state distribution of presidential disaster declarations under the Stafford Act for all types of disasters. All states except Idaho and Wyoming were declared at least once during the period January 1989 through April 1995. The highest incidence of declarations occurred in the Gulf Coast states and along the Pacific Coast, plus Arizona. Texas led the nation with 19 declarations, followed by California (15), Washington (14), Arizona (12), and Oregon (11). Each of the remaining states was declared between once and 10 times during the study period. The Atlantic Coast states (except Maine), despite high concentrations of population, experienced relatively fewer declarations than the Gulf of Mexico and Pacific Coasts or the upper Midwest.

The reasons for this contrast are not obvious. Some of the variation may stem from the distribution of specific types of disasters, as depicted in Figures 2-7 through 2-15. The West experiences earthquakes, wildfires, landslides, and floods. The Gulf Coast is struck by hurricanes, tornadoes, and inland floods, while the Midwest is subject to floods, tornadoes, and winter storms. The East

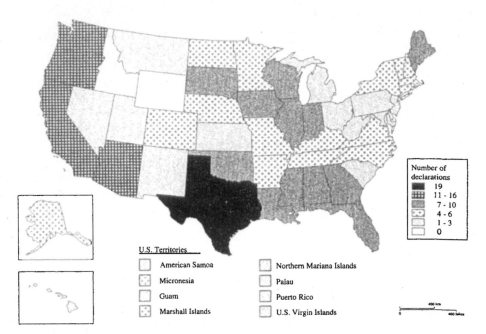

FIGURE 2-3 Total number of federal disaster declarations by state: January 1989 to April 1995. (Data source: Federal Emergency Management Agency. Graphics by Ute Dymon.)

Coast is prone to hurricanes, floods, and winter storms. The relative scarcity of hurricanes and floods along the East Coast during the study period may account for the relatively fewer disaster declarations for those states.

Another possible factor is the relative political influence of the southern and western states as compared with the east during the Bush and Clinton administrations. Also, western and southern states are much larger than most eastern states, and thus embrace more territory potentially subject to a disaster. The nation's east coast is shared by 14 states while the west coast, of comparable length, is divided among only three states.

A very different distribution appears in Figure 2-4 and Figure 2-5. Not only has California received a high number of declarations, it also swamps the rest of the nation in average federal cost per disaster, largely due to the Northridge Earthquake in January 1994 as well as the Loma Prieta Earthquake in 1989 (Figure 2-4). Florida, South Carolina, Puerto Rico, and the Virgin Islands comprise the next highest group of average disaster cost jurisdictions, due primarily to Hurricane Andrew in the case of Florida and Hurricane Hugo for the others.

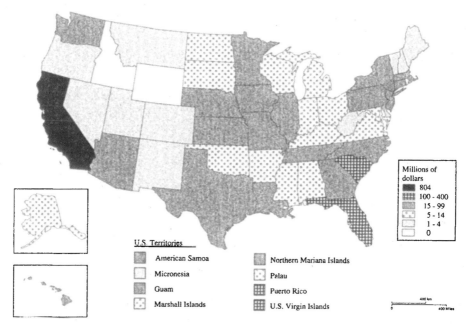

FIGURE 2-4 Average cost per Stafford Act disaster: January 1989 to April 1995. (Data source: Federal Emergency Management Agency. Graphics by Ute Dymon.)

Beyond these outliers, the rest of the states (except Idaho and Wyoming with no declarations) had average costs per disaster of less than $100 million. On the lower end, six sparsely populated western states, plus West Virginia, Delaware, Maine, New Hampshire, and Vermont had average federal costs per disaster of less than $5 million. At this level, one wonders why a presidential declaration was needed. It would seem well within the capabilities of any state to absorb $5 million in costs beyond what they otherwise are spending to recover from a disaster.

Figure 2-5 presents still another perspective on federal disaster assistance at the state level. In per capita payments, the high end is dominated by jurisdictions with comparatively small populations, such as Alaska, Hawaii, and the territories. American Samoa, the Northern Mariana Islands, and the U.S. Virgin Islands comprise an outlier group accounting for per capita assistance ranging from $1500 to $3600 during the study period. By contrast, all of the contiguous 48 states received federal disaster outlays of less than $150 per capita. The upper range of these states is occupied by California and Florida (despite their

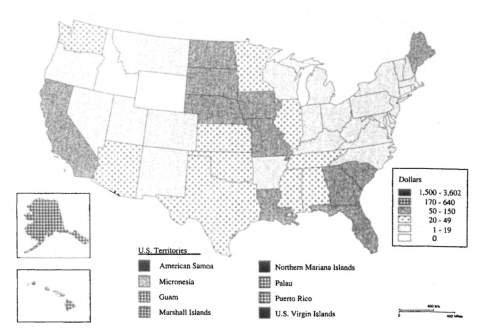

FIGURE 2-5 Per capita federal disaster assistance payments: January 1989 to
April 1995. (Data source: Federal Emergency Management Agency. Graphics by Ute Dymon.)

large populations) and by several low population states—Maine, the Dakotas,
and Nebraska. Missouri, Louisiana, Georgia, and South Carolina, with medium-
size populations, also belong to this group due to a plethora of floods during the
period in question.

The influence of very large disasters on allocations among states is again
reflected in Figure 2-6. California dwarfs the rest of the United States with
almost 60 percent of all federal disaster assistance under the Stafford Act. This
has been due primarily to the 1994 Northridge Earthquake, which accounted for
nearly one-half of all federal outlays (and over 10 times the federal costs resulting from the Loma Prieta Earthquake in 1989). California is in a class by itself
not only due to the very costly and devastating earthquakes but by being in the
high range in almost every disaster category including severe storms, fire suppression and fire disasters, and snow and ice disasters (see later figures). (Landslides and mudslides in 1998 became high-profile disasters as a result of the El
Niño storms along the Pacific coast, yet FEMA does not record those as distinct
categories of disaster.) The only type of disaster not declared in California during the study period was tornado.

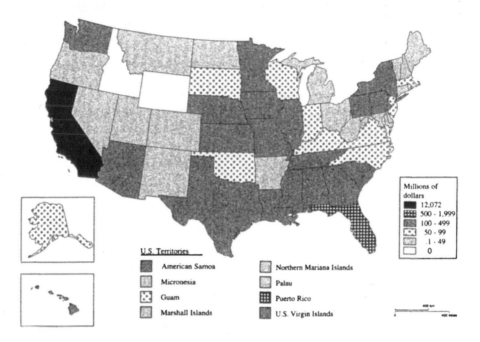

FIGURE 2-6 Total federal disaster assistance payments by state: January 1989 to April 1995. (Data source: Federal Emergency Management Agency. Graphics by Ute Dymon.)

The County Level

To gain a finer-grained picture of spatial patterns of disaster assistance, this study also mapped the frequency of disaster declarations at the county scale. Presidential disaster declarations are issued for individual states on a request from the governor of the affected state transmitted to the White House via the FEMA regional director, as supported by a preliminary damage assessment (PDA). A separate declaration is issued for each state affected by a multistate disaster; thus a single event may result in multiple declarations. Once a state is declared by the president, FEMA identifies the specific counties within that state that will be eligible for federal assistance.[10] Thus the initial determination of whether or not to declare a state is followed by a geographical determination as to which counties within that state will be eligible for Stafford Act benefits. That decision is based in part on the results of preliminary damage assessments and in part on politics and media attention. Individual counties may be designated to receive less than the entire suite of Stafford Act benefits (e.g., only public assistance or individual assistance may be authorized).

The mere inclusion of a county in a disaster declaration does not necessarily relate to the spatial distribution of Stafford Act expenditures—some counties may receive huge amounts and others very little (see Figure 2-7). Available resources did not allow the analysis of cost data at the county level. Nor does Figure 2-7 reveal what kind of disaster was involved in each declaration. But the map does not reflect the number of opportunities available to each county and its localities and inhabitants to receive Stafford Act assistance.

Access to disaster assistance is certainly not uniform or ubiquitous. Over one-third (1109 out of 3128) of all counties and county-type units were not listed in any disaster declaration during the study period and therefore received no Stafford Act assistance. (Flood insurance and other non-Stafford Act benefits of course were still available.) Another third (1079 counties) were designated once. The remainder (940 counties) were designated between two and seven times. The "disaster champions" were San Bernardino and Los Angeles counties in Southern California and Liberty County, Texas, just northwest of Houston. Those counties were designated either six or seven times during the six-year study period.

Some fairly distinct clusters of multiple designations are revealed by visual observation. The Pacific Coast has experienced repetitive county designations

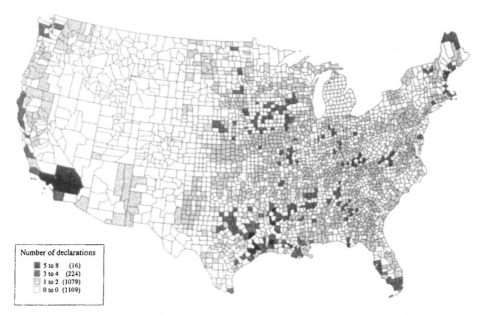

Number of declarations

- 5 to 8   (16)
- 3 to 4   (224)
- 1 to 2   (1079)
- 0 to 0   (1109)

FIGURE 2-7 Total disaster declarations by county: January 1989 to April 1995. (Data source: Federal Emergency Management Agency. Graphics by Ute Dymon.)

virtually from the Mexican border to the Canadian border. Concentrations of multiple designations clearly appear in Southern California (for earthquakes, wildfires, and floods) and bordering Puget Sound (floods). Another distinct cluster appears in the Lower Mississippi and Gulf Coastal region including eastern Texas, Louisiana, and northern Mississippi, where both coastal and inland flooding are frequent. Most Iowa counties have received multiple designations in contrast to neighboring counties in adjoining states such as Missouri, which has been relatively less active except for the very flood-prone St. Charles County at the junction of the Mississippi and the Missouri Rivers. This could reflect geographic differences in hazard risk between Iowa and its neighbors, or it may attest to the relative political influence of the states in question. For instance, ice jam flooding in the sparsely settled counties of northern Maine attracted frequent presidential declarations, likely reflecting the political influence of the former U.S. Senate Majority Leader from the Downeast state, George Mitchell.

Counties never declared as disaster areas during the study period dominate the western Great Plains, the Rocky Mountain region, and the Great Basin. This region is sparsely populated, but so is Maine. Presumably this pattern of inactivity simply reflects the lack of noteworthy floods, earthquakes, or other disasters except wildfires in this region. Counties in the Lower Peninsula of Michigan were largely undeclared in contrast to adjacent counties in Ohio and Indiana. (Does the "tornado belt" exclude Michigan?) Counties in the oft-flooded Red River Valley between the Dakotas and Minnesota experienced repeated declarations (including those of 1997 not included on this map). But further east, the lake-strewn wilderness of northern Minnesota was not declared, presumably because population centers were less threatened by riverine flood than along the Red River, and Minnesota lacked a Senator Mitchell. Elsewhere, there is no obvious explanation other than politics for observed anomalies. For instance, few counties were declared in West Virginia while neighboring counties in eastern Kentucky were declared frequently.

Thus the distribution of declaration activity is in part evident from the level of hazardousness for certain regions of the country. But comparing West Virginia with eastern Kentucky or northern Maine with northern Minnesota suggests that social and political factors also play a significant, and even a decisive role in determining which counties may be declared.

DISASTER GERRYMANDERING

The political dimension in the disaster declaration process was highlighted in the wake of the spring floods in Pennsylvania, where several major rivers crested and there was severe flood damage throughout the state. The *Pittsburgh Post-Gazette* reported on January 24, 1996, that the president declared six counties within the Susquehanna Valley near Harrisburg as disaster areas, bypassing the western part of the state around Pittsburgh, in spite of devastating floods experienced there. Governor Ridge blasted federal officials, saying that they should "stop playing games with Pennsylvania" and that Pennsylvania voters would remember this when it came to election time in the fall. Suddenly, 58 out of 87 Pennsylvania counties were declared as federal disaster areas. In addition, a nasty political game ensued in Congress in which Democratic and Republican senators pointed fingers at each other with accusations about who was participating in budget cuts for disaster relief (*Pittsburgh Post-Gazette,* January 24, 1996).

Similar "disaster gerrymandering" occurred in California after its widespread floods in January and February 1995. According to a "quick response" study by Richard T. Sylves,[11] two disaster declarations ensued a few weeks apart, ultimately including 57 out of 58 California counties. Moved by national media attention (the "CNN syndrome") and the political importance of California, President Clinton waived the requirement for a preliminary damage assessment in each county. The second declaration required only 40 minutes between the request by Governor Pete Wilson and the declaration by President Clinton.[12] According to Sylves: "Waiving standard declaration requirements allowed many counties to join the declaration without having to initially prove either need or recovery incapacity . . . localities have a tremendous incentive to seek a presidential disaster declaration if they experience damage in any form, even if [they are] able to recover from the event on their own."[13]

The promise of 75 percent federal funding after declared disasters and the rising probability of receiving a declaration for events of smaller magnitude than in the past may be disincentives for states and local communities to maintain and expand their own disaster-related capabilities. Indeed, the lower the nonfederal level of preparedness, the higher the likelihood that federal assistance may be approved. As noted by the National Performance Review in 1993:

> The system as a whole encourages state and local elected officials to ask for maximum federal disaster assistance. Requests have increased approximately 50 percent over the last 10 years. Even minor emergencies have been awarded full compensation. . . . To prevent the federal government from becoming the states' firstline resource in every emergency, objective criteria are needed to replace political factors affecting decisions about disaster declarations and ensuing response.[14]

Similarly, the House of Representatives Bipartisan Task Force Report in 1994 warned that federal overgenerosity in disaster assistance could undermine nonfederal disaster mitigation and preparedness efforts.[15] (The efforts of the Clinton administration since 1993 to incorporate mitigation into the federal emergency management process are considered in Chapter 3.)

The inexorable albeit uneven rise in federal disaster costs over time is related in part to the decision of Congress to establish a fixed federal cost share of 75 percent of eligible costs for all declared disasters. Such a fixed rate contradicts the principle that federal assistance is only intended to *supplement* emergency management capabilities of state and local governments. Since those capabilities would be expected to vary from one state and disaster to another, a more flexible federal cost share would better respect the concept that federal aid supplements, rather than displaces, nonfederal resources. Flexibility has been displayed only in cases where the president chooses to *increase* the federal share, for instance, to 90 percent for the Northridge Earthquake and to 100 percent of public assistance costs for Hurricane Andrew.

DECLARATIONS BY TYPE OF DISASTER

The remaining maps in this chapter depict the spatial distributions of specific types of disasters as identified by FEMA. There are some problems with FEMA's classification system where floods are concerned. In addition to declarations clearly labeled as "flood disasters" (Figure 2-8), FEMA also lists certain disasters as "hurricane and typhoon" (Figure 2-9), "tornado-flood combinations"

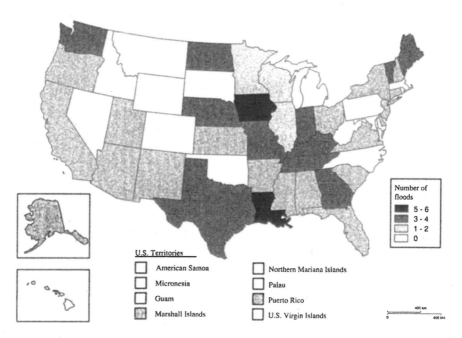

FIGURE 2-8 Flood disasters: January 1989 to April 1995. (Data source: Federal Emergency Management Agency. Graphics by Ute Dymon.)

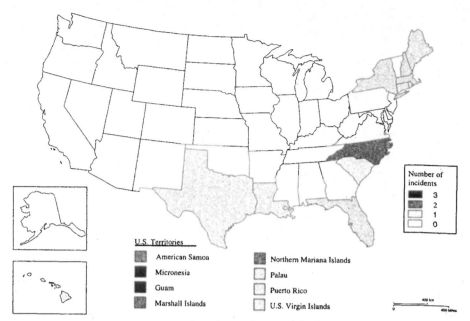

FIGURE 2-9 Hurricane and typhoon disaster declarations: January 1989 to April 1995. (Data source: Federal Emergency Management Agency. Graphics by Ute Dymon.)

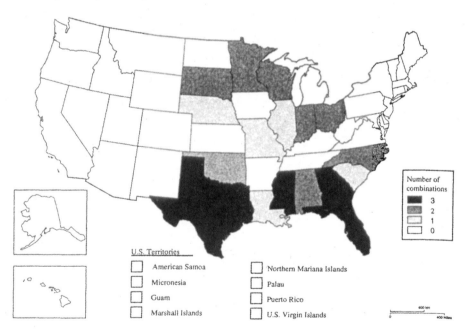

FIGURE 2-10 Tornado–flood combinations: January 1989 to April 1995. (Data source: Federal Emergency Management Agency. Graphics by Ute Dymon.)

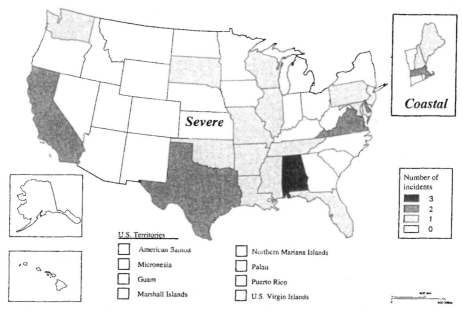

FIGURE 2-11 Non-hurricane coastal and other severe storm disasters: total declarations, January 1989 to April 1995. (Data source: Federal Emergency Management Agency. Graphics by Ute Dymon.)

(Figure 2-10), and "non-hurricane coastal and other severe storm disasters" (Figure 2-11).

About 90 percent of major disasters involve some form of flooding, including coastal storms and riverine disasters.[16] Over a 30-year period ending in 1994, average annual riverine flood damages exceeded $2 billion; over the last 10 years of that period, they exceeded $3 billion.[17] Among 231 major declarations issued between January 1989 and April 1995, about 145 were recorded by FEMA as involving floods, often in combination with other sources of damage.[18] The largest flood-related loss to date was the 1993 Midwest Flood, which accounted for about $4.2 billion in direct federal disaster expenditures, $1.3 billion in payments from federal insurance programs (primarily agricultural), and $621 million in federal loans to individuals businesses and communities.[19]

Flooding to some extent is involved in all these categories and yet it is impossible to break out federal expenses due to such flooding versus other effects such as wind. Although assignment to one category or another is somewhat arbitrary, no disaster is reflected on more than one map (double counted).

Fire suppressions and fire disasters account for the most declarations (Figure

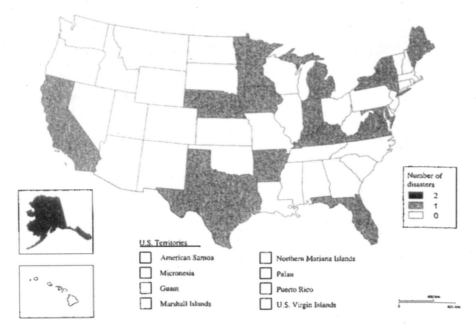

FIGURE 2-15 Snow and ice disasters: January 1989 to April 1995. (Data source: Federal Emergency Management Agency. Graphics by Ute Dymon.)

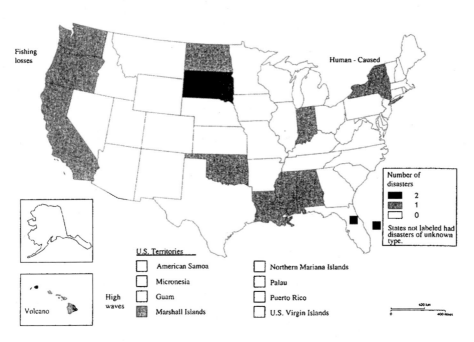

FIGURE 2-16 Miscellaneous disasters: January 1989 to April 1995. (Data source: Federal Emergency Management Agency. Graphics by Ute Dymon.)

disasters" (Figure 2-16). The terrorist attacks at the World Trade Center in 1993 and in Oklahoma City in 1995 were each declared major disasters, quite justifiably. But none was issued after the explosion of TWA Flight 800 in 1996 or any other aircraft disaster. The temptation to use disaster assistance to bail out ailing local or state economies in the absence of a natural disaster has generally been resisted so far. Exceptions were declarations issued to California, Oregon, Washington, and Massachusetts relating to fishing industry losses.

CONCLUSION

Chapter 2 has examined broad patterns of federal disaster assistance administered under the Stafford Act. A major finding of this analysis is the dominance of very large disasters and the states that experience them. In a sense, there is California and there is the rest of the United States in terms of total disaster funds allocated and average cost per disaster. Per capita allocations are also among the highest rank in the contiguous 48 states (although "offshore" portions of the nation received considerably higher per capita outlays). Much of California's dominance is attributable to the very costly Northridge Earthquake, which alone accounted for 48 percent of total Stafford Act outlays during the period of the study. But as documented by Sylves,[20] the White House has been particularly receptive to requests from California for declarations relating to other disasters such as the 1995 floods. All California counties except one were included in that declaration despite (or because of) a waiver by President Clinton of the need to conduct preliminary damage assessments in each county.

The five largest recipients of federal disaster assistance (California, Florida, South Carolina, Georgia, and Puerto Rico) accounted for three-fourths of all Stafford Act assistance during the study period. The rest of the United States thus received a total of about $5.5 billion, or about $1 billion per year. This is not necessarily an unjustified level of federal disaster assistance expenditure. However, it leaves two areas of concern: (1) Why are the smaller disasters amounting to $5 million or less in federal assistance declared by the president in the first place, and (2) how can the "budget busting" outlays for a mega-disaster such as the Northridge Earthquake or Hurricane Andrew be contained?

The first question presents the need for criteria to define the low-end threshold of disaster declarations. Too low a threshold reinforces the perception that the federal government will always come like the cavalry to rescue states and local governments from their improvident failure to prepare for routine disasters. Lapses in preparedness, response, recovery, and mitigation (to cite the disaster management litany) should not be encouraged by a too readily available bailout by the federal government and taxpayers.

The second question involves placing limits on federal largesse in disasters that unquestionably justify federal involvement. Again, states and local governments must provide what they reasonably can be expected to do for themselves. Some elements of disaster assistance (such as property tax losses) perhaps should be regarded as foreseeable local or state expenses, not to be reimbursed by the federal treasury, which after all does not indemnify against tax losses for nondisaster reasons such as the loss of a major employer. The 75:25 sharing of costs between federal and nonfederal entities is extremely generous to the latter, especially when the nonfederal share may be obtained from federal CDBG funds or in-kind services (expenses that would be incurred anyway). The reduction or waiver of the nonfederal share after Hurricane Andrew and the Northridge Earthquake were political decisions that contributed to the disproportionate share of Stafford Act funds allotted to those disasters. (In 1996, the federal share for Hurricane Fran in North Carolina was set at 90 percent.)

Aside from the largest disasters and most favored jurisdictions, the rest of the states and counties vary widely in the number of declarations and federal disaster dollars they have received. Some of this variance appears to relate to the geography of natural hazards, which are not evenly distributed in terms of risk and vulnerability. The population at risk may be another factor to be explored in future research. But the influence of politics and national media attention, as examined by Richard Sylves, clearly is a factor in many presidential declarations or in the denial of requests for declarations. Disaster assistance is an almost perfect political currency. It serves humanitarian purposes that only the cynical academic could question. It is largely funded out of supplemental appropriations and thus does not officially add to the budget deficit. It promotes the local economy of the area where the rebuilding process occurs. And it is extremely difficult to pinpoint exactly how much money the federal government spends on disasters. In the words of one FEMA staff member (who will remain anonymous), disaster assistance is the last big source of pork barrel in the federal government.

NOTES

1. Dr. Dymon was assisted in the preparation of the maps for this chapter by Frank Boscoe, a master's candidate in geography at Kent State University.

2. Ian Burton, Robert W. Kates, and Gilbert F. White (1993). *The Environment as Hazard*, 2d ed. New York: Guilford Press, Ch. 1. While the United States has not experienced a large number of natural disaster deaths and injuries in recent decades, elsewhere in the world there have been many very deadly disasters, for example: The Armenian Earthquake (1988)—about 55,000 fatalities; the Mexico City Earthquake (1985)—about 10,000 fatalities; and Bangladesh floods (1991)—about 140,000 persons killed.

3. National Academy for Public Administration (1993). *Coping with Catastrophe*. Washington, D.C.: NAPA; U.S. Senate Bipartisan Task Force on Funding Disaster Relief (1995). *Federal Disaster Assistance* Doc. 104-4 (104th Cong., 1st sess.) Washington, D.C.: U.S. Government Printing Office; U.S. House of Representatives Bipartisan Natural Disasters Task Force (1994). *Report*. Washington, D.C.: mimeo; U.S. General Account-

ing Office (1995). *Disaster Assistance: Information on Expenditures and Proposals to Improve Effectiveness and Reduce Future Costs.* (GAO-T-RCED-95-140). Washington, D.C.: GAO.

4. U.S. General Accounting Office (1996). *Disaster Assistance: Improvements Needed in Determining Eligibility for Public Assistance.* GAO/RCED-96-113. Washington, D.C.: GAO, Table 1.1.

5. Raymond J. Burby (1998). "Natural Hazards and Land Use: An Introduction," in R. J. Burby, ed., *Cooperating with Nature: Confronting Natural Hazards with Land-Use Planning for Sustainable Communities.* Washington, D.C.: Joseph Henry Press, 1-26, at 3.

6. See note 3.

7. Interagency Floodplain Management Review Committee (1994). *Sharing the Challenge: Floodplain Management into the 21st Century* (Galloway Report). Washington, D.C.: USGPO, 22.

8. To address this problem, the National Research Council formed a Committee on Assessing the Costs of Natural Disasters in December 1997. A parallel investigation is underway at this writing to investigate the full range of costs of coastal hazards under the auspices of the H. John Heinz III Center for Science, Economics, and the Environment.

9. Richard Roth, Calif. Dept. of Insurance. Remarks to NRC Committee on Assessing the Costs of Natural Disasters, Dec. 15, 1997.

10. U.S. General Accounting Office (1995). *Information on Declarations for Urban and Rural Areas* GAO/RCED-95-242. Washington, D.C.: GAO, 1.

11. Richard T. Sylves (no date). "California Flood Lessons." Quick Response Grant Final Report. mimeo.

12. Ibid., 9.

13. Ibid., 8.

14. National Performance Review (1993). *Creating a Government that Works Better and Costs Less: FEMA.* Washington, D.C.: Office of the Vice President, 13.

15. U.S. House of Representatives Bipartisan Natural Disasters Task Force (1994). *Report.* (Dec. 14) mimeo, p. 1.

16. Letter from Elaine McReynolds, Federal Insurance Administrator, to *The Wall Street Journal,* Sept. 15, 1995.

17. Galloway Report, note 7, xxi.

18. Some of these related to the same event, since a separate declaration is issued for each state significantly affected by a multistate disaster.

19. Galloway Report, note 7, 22.

20. Sylves, note 11.

CHAPTER 3

Stemming the Losses: The Quest for Hazard Mitigation

Rutherford H. Platt and Claire B. Rubin

> Building new Federal-State-local partnerships and public-private partnerships is the most effective means of implementing measures to reduce the impacts of natural hazards.
>
> FEMA, *National Mitigation Strategy,*
> December 6, 1995

> Extending insurance in the face of irresponsible placement of property, building infrastructure that storms are bound to wash away, and encouraging any policy that continues a cycle of predictable damage and guaranteed repair—all that sounds like the recipe for chronic disaster.
>
> Editorial: "Money Washed Away," *The News & Observer,*
> Raleigh, North Carolina, September 3, 1998, 14A
> (written after Hurricane Bonnie)

INTRODUCTION

The relentless rise of federal disaster assistance costs documented in Chapters 1 and 2 has long provoked demands for measures to anticipate and lessen the effects of extreme natural events. This has been especially the case with coastal and riverine floods that are relatively place-specific and that account for the bulk of disaster costs nationally in the average year (barring the occasional Northridge Earthquake). Flood hazard reduction has therefore served as the testing ground for ways and means of building safer and more sustainable communities through the concept of "hazard mitigation."[1]

Many researchers classify social response to natural disasters into four overlapping stages: (1) predisaster planning and preparation, (2) emergency response

69

and rescue, (3) short- and long-term recovery, and (4) mitigation[2] (Figure 3-1). The fourth phase, mitigation, feeds back into the first phase and contributes to predisaster planning for future disasters. Mitigation ideally occurs before a disaster occurs, but more commonly is addressed (if not fully achieved) in the process of recovery from disasters. The post-disaster recovery period is often referred to as a "window of opportunity" when political and media attention are riveted on the subject of flood, earthquake, hurricane, blizzard, or whatever event may have just happened. At that moment, the subject of hazard mitigation is most salient to the general public and therefore to the political process.[3] Learning the lessons of disaster is a common theme in political rhetoric and academic research in the aftermath of extreme natural events.

Mitigation has been defined as "any action taken to permanently eliminate or reduce the long-term risk to human life and property and the negative impacts on natural and cultural resources that can be caused by natural and technological hazards."[4] Before the 1970s, mitigation of flood hazards generally took the form of structural projects to restrain or deflect floodflows and coastal storm surges. The Ash Wednesday Nor'easter of March 1962, for instance, prompted legislative action to line the shores of the East Coast megalopolis with sea walls, groinfields, and beach nourishment projects under the direction of the Army Corps of Engineers.[5]

In more recent usage, the term *hazard mitigation* embraces a wide diversity of actions to reduce vulnerability other than structural measures, for example:

- Warning and evacuation planning
- Building regulations (for new construction, rebuilding, or retrofit of existing buildings)
- Land use planning and regulation
- Property acquisition
- Public information
- Protection and restoration of wetlands and other natural habitats

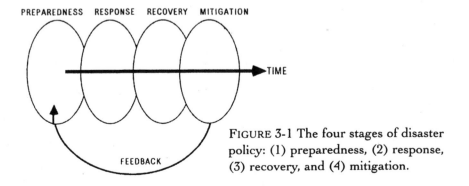

FIGURE 3-1 The four stages of disaster policy: (1) preparedness, (2) response, (3) recovery, and (4) mitigation.

For purposes of this chapter, the term will be used to refer to "nonstructural" measures such as those bulleted earlier. While the focus of this discussion will be on flood and coastal hazards, most of the actions listed apply to other types of natural hazards as well.

According to a 1994 report of the National Academy of Sciences: "Mitigation has been an underlying requirement of federal emergency management policy for about 30 years, beginning with floodplain management requirements in the 1960s. In actual practice, however, only a fraction of the mitigation measures known to be effective have been implemented."[6] An experienced mitigation researcher, Raymond J. Burby, in reflecting on the aftermath of the 1993 Midwest Floods, has written:

> Federal agencies have not effectively used existing knowledge on private-sector decisions related to hazard mitigation and, except for the National Science Foundation, have not known enough to invest in building knowledge about floodplain management that would enable them to deliver programs more effectively. As a result, some federal programs have not penetrated private markets adequately (flood insurance, for example . . . and many opportunities to foster private retrofitting . . . are lost due to the absence of information about how to act effectively.[7]

Mitigation has had a checkered history over the past three decades. While universally supported in principle, it has often proven to be the unwelcome guest at the post-disaster banquet. Rebuilding more safely may cost more, take longer, and sometimes conflicts with private property interests and public tax base and economic priorities. And despite recent expansion of funding for mitigation, the lion's share of federal disaster assistance is still devoted to rebuilding the *status quo ante,* as quickly as possible.

Under the Clinton administration, the Federal Emergency Management Agency (FEMA) has established a Mitigation Directorate, has issued a National Mitigation Strategy, and has launched several initiatives to stimulate community-level mitigation efforts (e.g., Project Impact in 1997). But these are merely the latest in a long series of mandates, policies, and approaches to reducing vulnerability to natural disasters. After three decades of the National Flood Insurance Program, two decades since the formation of FEMA; and over one decade since the Stafford Act, there has been no systematic effort to evaluate the effectiveness of various approaches to flood hazard mitigation, let alone mitigation of other natural risks. The federal government, and particularly FEMA, are still struggling to define, achieve, evaluate, and improve their efforts in hazard mitigation. Despite abundant rhetoric, it remains unclear what mitigation really means, and who should pay for it.

Further complicating the picture, governmental regulation of private land use has become politically unpopular during the 1990s. Although Chapter 5 argues that there is still a strong judicial basis for the use of noncompensatory land use and building regulations in hazardous areas, there is a tendency to eschew "command and control" (as regulation is now sometimes referred to) in favor of various forms of voluntary "partnering" to achieve environmental goals. Consistent with this broader trend, the rhetoric of hazard mitigation, as expressed in the National Mitigation Strategy and other policy documents, now focuses on *voluntary* joint efforts by FEMA, states, local communities, and the private sector to reduce vulnerability to natural disasters.[8]

The key issues addressed in this chapter and the case studies later in the book are: (1) How can vulnerability to natural hazards be reduced through wise land use practices? (2) How should the economic and political burdens of achieving wise land use practices be allocated among the federal government, the states, local communities, and private owners? and (3) Whether and how may the federal government require mitigation by nonfederal interests as a condition of receiving federal disaster assistance?

A different way to pose the issue is whether mitigation in the 1990s is largely a new form of federal disaster assistance funding or whether vulnerable states, communities, and property owners are accountable for reducing hazards within their respective areas of authority. If the former view prevails, nonstructural mitigation remains merely a light fiscal tap compared with the sledge hammer blow of disaster recovery funding.[9] If the latter view prevails, the scope of governmental regulation under the police power is increasingly limited in the face of the property rights movement and resistance by local government to state or federal intrusion in their land use policies and decisions. In other words, how can hazard mitigation be accomplished where private owners and local communities decline to be the responsible party or even to "partner" with the state and federal government?

THE SYNERGY OF DISASTERS AND MITIGATION

Societal efforts to mitigate the effects of natural hazards have evolved episodically in response to disasters themselves, improved perception of hazards, and risk assessments (Figure 3-2). Most coastal protection projects of the 1950s and 1960s were authorized by Congress in direct response to a particular storm or series of storms.[10] Since the 1970s, change in laws and policies has increasingly been influenced by research findings and independent evaluations by other federal agencies, which themselves were conducted in response to specific disasters. The evolution of federal mitigation policy therefore is best described as an interactive process of synergy between three types of milestones: (1) natural disasters, (2) policy assessments, and (3) legal and organizational changes:

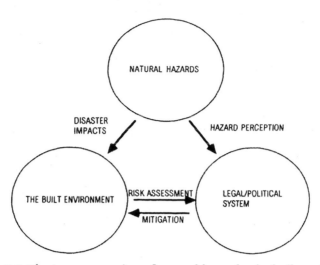

FIGURE 3-2 The interconnection of natural hazards, the built environment, and the legal/political system.

1. *Disaster events*—including major or catastrophic disaster events, which because of their size, problems, or consequences have become defining events for the emergency management community nationally
2. *Policy assessments*—these take many forms, e.g., special post-disaster evaluations, congressional studies and reports, and academic studies. Some of these have significantly influenced, either directly or indirectly, policy makers and legislators responsible for disaster policy.
3. *Laws and other actions*—including authority to build structural projects, creation and amendment of various financial assistance programs, executive orders, restructuring of governmental agencies

Figure 3-3 depicts the interaction of the three classes of milestones during the first two decades of active federal interest in nonstructural mitigation; more recent experience appears in Figure 3-4. Chronology, of course, does not always signify causality ("Post hoc, ergo propter hoc"). However, legislative history reveals numerous cases where disasters have been the impetus to statutory or other policy changes. Until the mid-1960s, congressional hearings were the primary linkage between a natural disaster and remedial legislation, as in the case of the Alaskan earthquake of 1964 and the Alaskan Earthquake Assistance Act (PL 88-451) passed later the same year. But with the adoption of the National Flood Insurance Act in 1968, the translation of disaster experience into legislative response has increasingly been intermediated by nonpolitical input in the

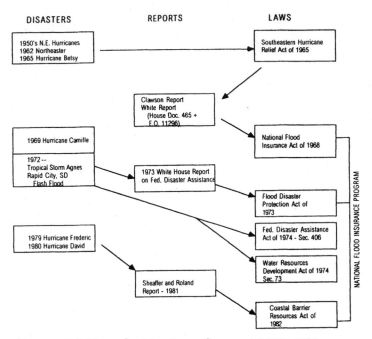

FIGURE 3-3 Hazard mitigation milestones: 1965–1982.

form of special "blue ribbon" panel reports, consultant studies, and academic research. This has in turn broadened the range of policy adjustments to natural hazards and, in particular, has strengthened the case for nonstructural hazard mitigation.

ORIGINS OF NONSTRUCTURAL MITIGATION (1965–1983)

The mid-Atlantic and New England states were battered by major hurricanes in the mid-1950s, and again by the Ash Wednesday Nor'easter in March 1962. As usual, Congress responded to these disasters with funding to restore seawalls, nourish beaches, and channelize inland streams. But the era of automatic reliance on structural projects was soon to end. During the 1950s and early 1960s, geographer Gilbert F. White and his colleagues and students at the University of Chicago were generating a "harvest of research" on the shortcomings of structural approaches to the flood problem.[11] In particular, they found that structural flood control and shore protection projects often induce a false sense of security leading to new investment in the area thought to be protected by the project. In the event of a flood or coastal storm exceeding the design capacity of the project, the ensuing losses on the new development were much larger than if that area had remained unprotected and relatively less developed. Instead of trying only to control floodwaters, the Chicago researchers argued, public policies should pursue a wider range of adjustments to also try to reduce the impact of floods on communities and individuals. These included such measures as land use planning and regulation, land acquisition, flood predictions and warning, and, under appropriate conditions, flood insurance.

Floodplain management was still rare in the 1950s,[12] but two federal agencies developed pilot programs under the stimulus of White's prodding. In 1953, the Tennessee Valley Authority (TVA) established a program of floodplain mapping and regulation within certain portions of its region. That served as a model for the Army Corps of Engineers flood hazard mapping program beginning in the early 1960s.

There was no model for the concept of government-sponsored flood insurance. For decades, the U.S. private insurance industry declined to offer flood insurance coverage due to the problem of adverse selection: The only people interested in it tend to be located in flood hazard areas and thus have a high risk of loss. The price of such coverage accordingly would have to be so high as to be unaffordable by most potential purchasers. It was therefore proposed in the early 1950s that the federal government should offer flood insurance coverage at low rates subsidized by the taxpayer. Congress in 1956 passed a national flood insurance act. But in the absence of funding, it was stillborn.

When Hurricane Betsy struck the Gulf Coast in 1965, Congress decided to take a fresh look at the nation's flood policies. The Southeastern Hurricane Relief Act passed that year directed the newly established Department of Housing and Urban Development (HUD) to "Undertake an immediate study of alternative programs which could be established to help provide financial assistance to those suffering property losses in flood and other natural disasters, including alternative methods of federal disaster insurance. . . ."[13] The ensuing study was conducted for HUD by resource economist Marion Clawson.

At about the same time, the Bureau of the Budget formed a Task Force on Federal Flood Control Policy, chaired by Gilbert F. White. The two reports jointly defined a turning point in national flood policy and hazard mitigation. Both concurred that federal flood control works and disaster assistance, without limits on new construction in vulnerable areas, were encouraging occupance of floodplains and rising flood losses. Both expressed cautious optimism regarding the prospects for a National Flood Insurance Program provided it incorporated measures to deter further unwise construction in floodplains. These would involve, among other elements, federal standards for local floodplain management. According to Clawson:

> It will be incumbent on the federal flood insurance agency to develop criteria for land use management of floodprone areas. Such criteria should be comprehensive and flexible enough to meet varying local conditions. . . . Every encouragement and

incentive should be given to local authorities to adopt and enforce
land use regulations in conformity with the national criteria.[14]

The Bureau of the Budget report[15] drew heavily on White's own research
and the experience of other members of the task force including James Good-
ard, director of the TVA floodplain management program. The task force criticized federal policies that promote unwise and hazardous development in floodplains:

> Despite substantial efforts, flood losses are mounting and *uneco-
> nomic uses of the nation's flood plains are inadvertently encouraged.* The
> country is faced with a continuing sequence of losses, protection,
> and more losses. While flood protection of existing property
> should receive public support, supplemental measures should
> assure that future developments in the floodplains yield benefits
> in excess of their costs to the nation.[16]

The report called for:

> . . . a new set of initiatives by established federal agencies with
> the aid of state agencies to stimulate and support sound planning
> at the local government and citizen level. . . . Planning and coor-
> dinating the development of the flood plain is required as part of
> any significant effort to break the pattern being fostered by pre-
> sent federal policies concerning flood damage prevention, namely
> the continuing sequence of losses, protection, and more losses.[17]

To break this cycle, the task force urged that floodplain mapping by such federal agencies as the Army Corps of Engineers, the Soil Conservation Service, and
the U.S. Geological Survey was needed to support "reasonable controls over
development of floodplains by local, regional, or state authorities."[18] Such regulations must be "based on sound, impartial definition of the facts if they are to
stand up in court as reasonable and fair."[19] Furthermore, federal agencies should
provide "limited technical assistance and encouragement" to state and local
planners and officials and individuals in the preparation of floodplain regulation
and the application of flood data for assessing floodplain location.[20]

The task force was cautious about flood insurance:

> A flood insurance program is a tool that should be used expertly
> or not at all. Correctly applied, it could promote wise use of
> floodplains. Incorrectly applied, it could exacerbate the whole

problem of flood losses. . . . It would not be improper to subsidize flood loss insurance for existing property. That might be done, provided owners of submarginal development were precluded from rebuilding destroyed or obsolete structures on the flood plain. However, to the extent that insurance were used to subsidize new capital investment, it would aggravate flood damages and constitute gross public irresponsibility.[21]

President Lyndon B. Johnson transmitted the task force report to Congress in August 1966 together with Executive Order 11926 (presumably ghostwritten by Gilbert White) lending presidential approval to nonstructural hazard mitigation:

To hold the nation's toll of flood losses in check and to promote wise use of its valley lands requires new and imaginative action. This nation's annual flood damage bill of more than $1 billion per year is excessive, even in a growing economy. Beyond the dollar loss the accompanying toll in personal hardship cannot be calculated. . . . *The key to resolving the problem lies above all else, in the intelligent planning for and state and local regulation of use of lands exposed to flood hazard*[22] (emphasis added).

In testimony to Congress in 1967, Under Secretary of Housing and Urban Development Robert C. Wood reaffirmed the linkage between flood insurance and mitigation: "It would not be logical as a matter of public policy to permit insurance to be made available in localities which did not, on their own initiative, or on the initiative of state authorities, take whatever steps would be appropriate to assure that their citizens would not unknowingly acquire and develop property where it is subject to known flood hazards."[23]

The National Flood Insurance Program, as established by Congress in 1968, incorporated at least four forms of mitigation:

1. Identification and mapping of 100-year floodplains in the nation's communities (Sec. 1360)
2. Establishment of minimum federal standards for new construction in mapped floodplains to be adopted by communities as a condition of participation in the NFIP (Sec. 1361)
3. Charging of actuarial rates for flood insurance coverage for new or substantially improved structures (as per the White and Clawson reports)
4. Authority for public purchase of chronically floodprone properties (Sec. 1362)

Regarding land use management in particular, Section 1361 of the act authorized the administrator of the National Flood Insurance Program to develop criteria for state and local adoption to:

1. Constrict the development of land that is exposed to flood damage where appropriate,
2. Guide the development of proposed construction away from locations that are threatened by flood hazards,
3. Assist in reducing damage caused by floods, and
4. Otherwise improve the long-range management and use of floodprone areas

The NFIP attracted little interest initially, due to delays in the preparation of flood hazard maps and the lack of buyer interest. In 1969, Congress permitted interim coverage to be made available pending completion of flood insurance studies and maps for local communities. But still the program languished in obscurity.

In 1972, two milestone flood disasters, Tropical Storm Agnes and the lethal flash flood at Rapid City, South Dakota, galvanized renewed presidential interest in disaster policy. A 1973 White House policy paper[24] stimulated Congress to adopt the Flood Disaster Protection Act to strengthen the NFIP and specifically to "require states or local communities, as a condition of future federal financial assistance, to participate in the flood insurance program and to adopt adequate flood plain ordinances with effective enforcement provisions consistent with federal standards to reduce or avoid future flood losses. . . ."[25] The act required that anyone borrowing from a federally related lender to buy or build on floodprone property must purchase a flood insurance policy. Also it limited disaster assistance in communities not participating in the NFIP. Community enrollment in the NFIP and sales of flood insurance both picked up. (See Table 1-6.) The 1973 Flood Disaster Protection Act also added flood-related erosion and mudslides as insurable hazards under the NFIP in response to recent experience with those hazards in several states.

The Disaster Relief Act of 1974 (93-288), itself another response to the disasters of 1972, included the first explicit congressional requirement for hazard mitigation as a prerequisite for federal disaster assistance. Section 406 of that law, which remains in effect today as Section 409 of the 1988 Stafford Act, provided that state and local governments receiving aid were required to "agree that the natural hazards in the areas in which the proceeds of the grants or loans are to be used shall be evaluated and appropriate action shall be taken to mitigate such hazards, *including safe land use and construction practices,* in accordance with [federal] standards. . . ."[26]

A third post-Agnes enactment, the Water Resources Development Act of

1974,[27] provided in Section 73 that nonstructural approaches to flood loss reduction should receive equal consideration with structural measures. The same act authorized the Corps of Engineers to acquire several thousand acres of wetlands in the Charles River Watershed near Boston to protect "natural valley flood storage" instead of building structural flood control projects.

Recovery from the flood disasters at Rapid City in 1972, at Big Thompson Canyon, Colorado, in 1976, and elsewhere was reviewed by the U.S. Water Resources Council (WRC) in a 1979 study of nonstructural flood mitigation.[28] WRC urged the creative use of federal disaster assistance to achieve significant reduction of vulnerability. Stressing the need to identify mitigation opportunities immediately after a major flood, it recommended the formation of mitigation teams to prepare a mitigation report within days after a disaster declaration. This proposal was adopted by the federal Office of Management and Budget in a 1980 directive, which in turn led to an interagency agreement to establish such teams. The agreement further provided that federal agencies would "conform their recovery actions to the recommendations of the report to the fullest extent practicable."[29]

With respect to seismic hazards, the United States enjoyed a period of relative freedom from major earthquakes during the 1960s and 1970s, except for the Alaska Earthquake of 1964 and the San Fernando Valley Earthquake of 1971. In contrast to floods, Congress took a longer-range view of earthquake preparedness. The National Earthquake Hazards Reduction Act of 1977[30] established a program of research, technical assistance, and public education to be administered jointly by FEMA, the U.S. Geological Survey, the National Science Foundation, and the National Institute of Standards and Technology. But unlike the NFIP, this act lacked the action-forcing provisions to promote the adoption of local mitigation measures. It offers neither the carrot of federal insurance nor the stick of federal standards. It simply called for "the development, publication, and promotion, in conjunction with state and local officials and professional organizations, of model building codes and other means to encourage consideration of information about seismic risk in making decisions about land use policy and construction activity. . . ."[31] (The absence of federal standards would be criticized in a 1993 report to Congress by a panel of experts convened to review the National Earthquake Hazards Reduction Program (NEHRP), as discussed below.)

President Jimmy Carter established the Federal Emergency Management Agency (FEMA) by executive order in 1979, stating that hazard mitigation would be a major focus of the new agency: ". . . Federal hazard mitigation activities should be closely linked with emergency preparedness and response functions. This reorganization would permit more rational decisions of the relative costs and benefits of alternative approaches to disasters. . . ."[32]

MITIGATION UNDER THE REAGAN/BUSH ADMINISTRATIONS (1983–1993)

The Coastal Barrier Resources Act

In 1979, Hurricane Frederic inflicted heavy damage on the Gulf Coast, particularly in the vicinity of Gulf Shores and Dauphin Island, Alabama. A study commissioned by FEMA documented that the federal government had originally encouraged development on such coastal barriers through flood insurance, sewer and water grants, beach nourishment, and other development incentives. It then paid much of the cost of ensuring damage through disaster assistance. Based on federal expenditures due to Hurricane Frederic in 1979, Sheaffer and Roland estimated that it would be cheaper for the federal government to buy the remaining undeveloped coastal barriers than to subsidize their development and then bear the costs of rebuilding them after a disaster.[33] This concern was reinforced by a 1982 GAO report that found federal flood insurance to be at least a "marginal added incentive for development in coastal and barrier island communities."[34] And a 1981 position paper by "Concerned Coastal Geologists" urged retreat from eroding shorelines in place of government-sponsored efforts to stabilize coasts and beaches.[35]

These various reports inspired Congress to adopt two measures to limit the availability of federal incentives to development on coastal barriers. In August 1981, the Omnibus Budget Reconciliation Act prohibited new flood insurance

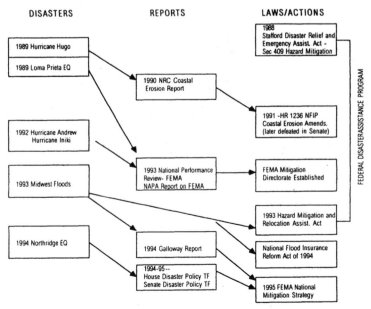

FIGURE 3-4 Hazard mitigation milestones: 1988–1995.

coverage for "any new construction or substantial improvements of structures located on undeveloped coastal barriers."[36] It applied this limitation to barriers that were neither substantially developed nor protected by a public agency or private conservation organization, and it directed the Department of the Interior (DOI) to identify and map all areas along the Atlantic and Gulf Coasts that met this definition. The resulting Barrier Island Inventory identified 186 sites involving about 640,000 acres of beach, dune, and wetlands.

Fourteen months later, Congress adopted the Coastal Barrier Resources Act of 1982,[37] which expanded the withdrawal of federal development incentives for undeveloped and unprotected coastal barriers. The law expressed the concern of Congress, as enlightened by the Sheaffer and Roland study, that "Certain actions and programs of the federal government have subsidized and permitted development on coastal barriers and the result has been the loss of barrier resources, threats to human life, health, and property, and the expenditure of millions of tax dollars each year."[38] The act declared the 186 segments identified by DOI as the "Coastal Barrier Resources System" (CBRS). Within that system, it reaffirmed the prohibition of new flood insurance coverage, and further limited federal financial assistance in such areas for the "construction of any structure, appurtenance, facility, or related infrastructure,"[39] including sewer and water systems, roads and causeways, and shore protection. Congress was thus persuaded that flood insurance and other federal incentives were inducing development in hazardous and ecologically fragile coastal barriers that ultimately would lead to further federal costs when those sites were struck by coastal disasters.

The Coastal Barrier Resources Act represented the "ideal case" of legislation embracing a new concept of mitigation, as adopted in response to policy recommendations based on government and academic research. The act was amended in 1990 to expand the CBRS to embrace a much wider list of coastal units, including some on the West Coast and the Great Lakes.[40] But the act remained applicable only to areas not yet developed or developing and thus did not curtail federal benefits to coastal barriers that are already partially or fully developed. Such "Cities on the Beach"[41] and nearby mainland coastal areas today are covered by hundreds of billions of dollars in federal flood insurance, as well as private casualty insurance, and represent an immense potential federal disaster assistance cost when they are hit by major hurricanes.[42] Federal flood insurance, beach nourishment, and other federal incentives to further development remain available in coastal areas exempted from the Coastal Barrier Resources Act. Mitigation of hazards in those areas thus remains a major challenge in the 1990s.

Floodplain Management Assessment

In 1980, a Federal Interagency Task Force on Floodplain Management under the aegis of FEMA was established to confer and collaborate on national flood policy issues. In 1987, the Task Force commissioned a broad assessment of floodplain management, the first since the 1966 report of the Task Force on Federal

Flood Control Policy. The final report and summary volume issued in 1992[43] comprised "a compilation of available information concerning the nation's flood-plains, experience with tools and strategies to reduce losses of life, property and environmental resources, and a perspective of what has been accomplished."[44] As a product of an interagency task force, however, the assessment overall tended to ratify the status quo and to support the agencies in what they were already doing. A more pessimistic appraisal emerged in a 1989 critique of the draft Assessment report by an ad hoc National Review Committee chaired by Gilbert F. White:[45]

> The present status of floodplain management does not encourage complacency. The record is mixed. There are encouraging trends, as with the number of communities having some form of flood-plain regulations, but the rising toll of average annual flood losses has not been stopped or reversed. Some activities look more pro-ductive on paper than on the ground or in the real vulnerability of people. On balance, progress has been far short of what is desir-able or possible, or what was envisaged at times when the current policies and activities were initiated.
>
> . . . Losses to the nation from occupance of riverine and coastal areas subject to inundation are continuing to escalate in constant dollars. Some of the losses can be attributed to failure to complete protection works. Most can be attributed to increased property at risk.[46]

This appraisal was based on the collective experience of the committee mem-bers and their chair, not on empirical investigation of the actual effects of the National Flood Insurance Program on actual trends of riverine and coastal losses. To date, no such evaluation has been sought by FEMA or undertaken nationally.

Adequacy of Federal Disaster Response

In 1988, Congress passed the Stafford Disaster Relief and Emergency Assistance Act (Stafford Act),[47] amending the 1974 Disaster Relief Act and prior laws dat-ing back to 1950. Section 409 of the Stafford Act reiterated the long ignored mandate of the 1974 law that state and local governments receiving federal dis-aster assistance must assess and implement opportunities to reduce natural haz-ard risks.

> Sec. 409. As a condition of any disaster loan or grant made under the provisions of this act, the recipient shall agree that the nat-ural hazards in the areas in which the proceeds of grants are to be

used shall be evaluated and appropriate action shall be taken to mitigate such hazards, including safe land use and construction practices. . . .[48]

Section 404 of the Stafford Act created a new hazard mitigation grant program to assist states in conducting projects identified by Section 409 plans that are "cost effective and which substantially reduce the risk of future damage, hardship, loss, or suffering in any area affected by a major disaster."[49] Up to 10 percent of federal public assistance funds pursuant to a disaster declaration may be spent on hazard mitigation projects (raised to 15 percent in 1993). The federal cost share of such projects was set at 50 percent by the Stafford Act (later raised to 75 percent).

In 1996, researchers at the University of North Carolina reviewed national experience with Section 409 hazard mitigation plans.[50] They found that for the most part, states have prepared some kind of hazard mitigation plan in compliance with the Stafford Act, but few of the plans are directly related to future 404 projects.

> The state 409 plans are at their strongest in describing and assessing present hazards, but they are weaker in assessing vulnerability of people and property to those hazards and in assessing "risk" in any systematic manner. . . . Mitigation action proposals tend to stress measures that are easiest to implement (promoting awareness, technical assistance) rather than actions that are more effective but require greater political commitment, cost more, and intervene more directly in the development process.[51]

Delivering Federal Assistance

Two milestone disasters of 1989, Hurricane Hugo and the Loma Prieta Earthquake, provoked a storm of political and media criticism regarding the adequacy of federal response to major disasters. In the process, issues concerning longer-term mitigation of vulnerability were temporarily eclipsed.

Hurricane Hugo resulted in four presidential declarations for the U.S. Virgin Islands, Puerto Rico, South Carolina, and North Carolina. It devastated St. Croix in the Virgin Islands and inflicted widespread coastal and inland damage in the Carolinas. Federal outlays exceeded $1 billion and total estimated damage was about $7 billion, which at that time was the most expensive disaster to date. The private insurance industry absorbed $4.28 billion in claims, a record at the time. (This, however, would be soon be surpassed by Hurricane Andrew with $15.5 billion in private insurance losses and the Northridge Earthquake with $11.2 billion.) In both the Virgin Islands and in South Carolina there were problems with the federal response, which showed up more significantly after Hurricane Andrew in Florida in 1992.

The Loma Prieta Earthquake (1989) struck the Bay Area of California just a few weeks after Hurricane Hugo. It was a moderate-sized earthquake (7.1 on the Richter Scale), but it was the largest earthquake to strike Northern California since 1906. The Bay Area had become densely urbanized, often with little regard for seismic hazards as in the filling and development of shorelands bordering the Bay. Much damage occurred to freeways and unreinforced masonry buildings on filled lands in San Francisco and Oakland, dozens of miles from the epicenter. The disaster was highly publicized since it disrupted the opening of a World Series game in Candlestick Park in San Francisco and was widely televised. Some special problems occurred in response programs, such as high impact on minority communities and problems with persons who were homeless before the earthquake seeking sheltering assistance. The estimated cost of damage was about $8 billion of which federal assistance accounted for about $1 billion. (See Chapter 8.)

Questions concerning the timeliness, efficiency, and coordination of federal response to Hurricane Hugo and Loma Prieta prompted an investigation by the U.S. General Accounting Office (GAO) in 1990. In its report,[52] GAO noted that preparedness problems are linked to ". . . weaknesses in state and local programs, as well as in FEMA's assistance and overall guidance."[53] Most of the specific problems cited were related to response and recovery activities, not to mitigation of future losses. The weaknesses in the federal response to these two events seem to be precursors of those that emerged after Hurricane Andrew in Florida, as documented by GAO in 1992 and 1993.

Reviewing the NEHRP

In 1990, the year after Loma Prieta, Congress reauthorized the National Earthquake Hazards Reduction Act,[54] adding language that called for seismological risk assessments to support "virtually every aspect of earthquake hazards management, whether emergency planning, public regulation, detailed building design, insurance rating, or investment decision [sic].[55] Mitigation of seismic hazards is also reflected in the statement that "The vulnerability of buildings, lifelines, public works . . . can be reduced through proper earthquake resistant design and construction practices."[56] The act also charged FEMA with preparing and recommending to Congress a plan for lifeline design and construction standards to be implemented pursuant to federal regulation.[57] Concurrently with this legislation, President Bush issued Executive Order 12699 requiring that federally owned or leased buildings be designed and constructed in accordance with appropriate seismic principles.

Since the advent of the program, the structural engineering community has quietly but effectively changed the national model building codes to incorporate seismic resistance standards for new construction. Under the auspices of the NEHRP office at FEMA, this was done on a voluntary basis and has been effective although not pervasive through all states with seismic risk. FEMA-funded efforts provide the resource documents in the form of provisions for new buildings.[58] These provided technical criteria (but did not mandate them) that were subsequently incorporated into the three major model building codes used in the United States. In turn, some at-risk states and localities have adopted or adapted these model codes for their own use.

However, the general approach and efficacy of NEHRP was strongly questioned in two reports issued in January 1993. The report of the NEHRP Advisory Committee, established by the 1990 Act, warned that the NEHRP suffered from lack of unified administrative authority and an inability to elicit seismic risk mitigation efforts uniformly among states and local governments.

> NEHRP has had limited success in achieving the nationwide adoption and enforcement of much of the technology that it helped develop, although the Program has devoted measurable effort to stimulating such application of risk-reduction techniques.

The committee, chaired by former Federal Insurance Administrator George K. Bernstein, urged that "the most crucial part of the implementation of earthquake risk reduction should be the identification and enactment of a series of financial incentives and mandates to encourage state and local governments to enact and enforce risk-reduction laws and codes, and to stimulate additional risk-reduction efforts within the private sector." The report suggested that such incentives might include:

- Requiring seismic risk reduction in all new construction funded or guaranteed by the federal government
- Federal income tax credits for seismic mitigation
- A federal matching grant for retrofit of critical state and local government facilities
- Requiring seismic mitigation as a condition of federal disaster assistance
- A federal disaster insurance program with a strong risk reduction requirement as the primary focus

The Committee concluded that "unless NEHRP legislation incorporates a programmatic implementation mechanism that creates strong incentives for the adoption of earthquake risk measures, the loss mitigation goals of the Program will not be achieved in the foreseeable future."[59]

A parallel report by a "panel of experts" (including several members of the NEHRP Advisory Committee) further criticized the failure of the program to stimulate adequate state and local effort to mitigate seismic hazards. In his issue paper, George Bernstein observed that two objectives of the NEHRP Act— (1) improved design and construction methods and practices and (2) land use controls and redevelopment—have not "been implemented broadly enough to hold out promise for comprehensive hazard reduction throughout the United States. . . . In the vast majority of the United States (including the most earthquake-prone areas), there is limited enforcement of seismic building code provisions."[60] Based on the precedent of the National Flood Insurance Program, Bernstein called for the federal government to "develop model earthquake codes that would be required to be adapted, adopted, and implemented at state and local levels based on state and local evaluation of the particular hazards involved."[61]

Hurricane Andrew and Its Aftermath

The final disaster milestone of the Reagan–Bush era was Hurricane Andrew, the nation's most costly hurricane to date, which struck Dade County, Florida, and portions of Louisiana in late August 1992. Hurricane Andrew destroyed over 28,000 homes and damaged approximately 107,380 others, resulting in the displacement of about 180,000 people from their homes and 75,000 from their jobs[62] (Figures 3-5 and 3-6). The combined costs incurred by private insurance companies ($15.5 billion on 680,000 claims) and federal disaster assistance ($2 billion) made it the worst disaster in economic terms to strike the United States. Much of the damage to homes and commercial buildings was caused by failure of roofs, windows, and doors and resulting rain damage to interior walls and contents. It has been estimated that one-quarter of the losses were caused by shoddy workmanship and poor enforcement of buildings codes.[63] Hurricane Andrew marked the first opportunity to put into operation the new Federal Response Plan involving 26 federal agencies and the American Red Cross. The plan was developed initially to deal with catastrophic earthquakes and later broadened in light of the deficiencies of federal response to Hurricane Hugo.

Two weeks after Hurricane Andrew, Hurricane Iniki struck Hawaii, causing intense devastation on the island of Kauai. No direct legislative response ensued from the Andrew/Iniki disasters since the Bush administration would soon leave office. In the midst of his reelection campaign, however, President Bush enlarged the federal share of disaster assistance in both recovery efforts to virtually 100 percent.

Criticism of federal response to Andrew once again prompted a series of policy and program reviews that would strongly influence the Clinton administration, which took office in January 1993.

FIGURE 3-5 Effects of Hurricane Andrew. Ground view of the devastation in the Pinewoods Villa. (National Hurricane Center)

Figure 3-6 Hurricane Andrew. Note the two vehicles still in the garage but overturned by the force of wind. (National Hurricane Center)

flood was later estimated at $12 to $16 billion, of which federal costs of response amounted to about $4.2 billion. About 100,000 housing units experienced flood damage and crop damage was very widespread.[73] (See Chapter 7 and Figure 3-4.)

Two institutional milestones soon followed this epic disaster. First, at the urging of the Clinton administration, Congress enacted the Hazard Mitigation and Relocation Assistance Act of 1993 (PL 103-181), which amended the Stafford Act to (a) raise the federal share of the cost of eligible mitigation measures from 50 percent to 75 percent and (b) raise the eligible funds available for mitigation projects from 10 percent to 15 percent of federal disaster assistance. As discussed in Chapter 6, FEMA applied this new authority to acquire some 8000 homes of flood victims, converting their building lots to public open space.

The second institutional milestone was the appointment by the White House of a special Interagency Floodplain Management Review Committee to assess the nature and magnitude of losses and to recommend changes in federal policies to reduce future losses. The committee study was directed by General Gerald E. Galloway, Jr., Dean of the Faculty and Professor of Geography at the U.S. Military Academy. The committee included a total of 31 professionals, 15 on loan from federal agencies in the Washington, D.C. area, and 16 members of a Scientific Assessment and Strategy Team (SAST) at the Earth Resources Observation System center (EROS) at Sioux Falls, South Dakota. The report was completed between January and June 1994. The research process was highly interactive. Working through the governors' offices of all nine affected states, meetings were held with state and local officials in over 60 locations. Extensive contacts were established with federal agencies and members of Congress and their staff in Washington. Concurrently, the SAST group in Sioux Falls generated a variety of computer analyses and images representing a multitude of geographic data.

The committee's report, *Sharing the Challenge* (known as the "Galloway Report"), struck a balance between engineering, agricultural, and environmental issues. The report praised the effectiveness of existing Corps of Engineers flood control projects in reducing damages, while calling for limits on the replacement or expansion of nonfederal levees that widely failed. The report called for the restoration of natural floodplain/wetland functions and abandonment of agriculture where feasible.

The Galloway Report also examined the mitigation of losses through flood insurance. In this, the largest inland flood since the advent of the National Flood Insurance Program in 1968, flood insurance played a very minor role. Total claims amounted to $293 million on 16,167 claims. This was the third largest payout under the NFIP, but it comprised only about 2 percent of the total damage estimate of $12 billion and about 14 percent of total federal disaster outlays. The report criticized the extent of NFIP market penetration.

The NFIP has not achieved the public participation needed to reach its objectives. This situation is evidenced by the assistance provided to individuals and businesses during the Midwest Floods. . . . Estimates of those covered by flood insurance nationwide range from 20 to 30 percent of the insurable buildings in identified flood hazard areas. Initial estimates in the Midwest flood area ranged from below 10 percent to up to 20 percent.[74]

On the subject of land use control in floodplains, the Galloway Report was politically cautious: "Land use control . . . is the sole responsibility of state, tribal, and local entities. . . . The federal responsibility rests with providing leadership, technical information, data, and advice to assist the states. . . ."[75]

Like the National Performance Review, the Galloway Report warned against allowing federal disaster assistance policy to create a "moral hazard"—lessening individual incentives to guard against disaster through overzealous government assistance programs. From FY 1989 and FY 1993, over $27.6 billion was spent on federal disaster assistance,[76] much of it through unbudgeted special appropriations. In three highly politicized recent disasters,[77] the federal share of disaster assistance was increased to virtually 100 percent from its previous level of 75 percent. The report cautioned against abandoning a meaningful nonfederal cost share and sense of personal and community responsibility in responding to disasters.

The House and Senate Task Force Reports

Soon after the Northridge Earthquake in January 1994, both Houses of Congress established bipartisan task forces to review federal disaster assistance policies, as discussed in Chapter 1. Both reports concurred that the availability of federal assistance has encouraged states, local governments, and the private sector to avoid responsibility for disaster preparedness and mitigation. But they differed considerably in their proposed remedies.

The House report focused primarily on financial incentives to reduce the burden on federal taxpayers by expanding individual responsibility for disaster losses. Listed first among its 54 recommendations was the establishment of a "private, nationally based all-hazard insurance program, in consultation with the insurance industry . . . for residential and commercial property."[78] Availability of insurance under this program would be contingent on conformity with "minimum natural disaster hazard mitigation portions of the model building codes, fire, life-safety, and siting ordinances, and make [sic] other mitigation efforts."[78] Rep. Joseph P. Kennedy II, a member of the task force, dissented from this recommendation, citing testimony by the General Accounting Office that all-hazard insurance could expose federal taxpayers "to a contingent liability of unknown, and perhaps massive, amounts."[80] GAO had warned that actuarial rates would be difficult to set and that insurance companies would assume the safest risks and pass on the least insurable risks to the federally established program (the moral hazard problem again).[81]

The Senate Task Force report provided a considerably more detailed analysis of federal disaster programs, funding levels, and issues. It estimated that the federal government had obligated almost $120 billion in direct grants, low interest loans, and insurance payments under 578 presidential disaster declarations between 1977 and 1993 (including federal expenditures for structural flood control projects).

Among many factors contributing to the rising costs of federal disaster assistance addressed by the Senate Task Force are (1) the increasing use of nonbudgeted "supplemental appropriations" (totaling $22 billion between 1992 and 1994) to fund specific disaster costs; (2) the waiver of the 25 percent nonfederal cost share in about 15 disasters since 1985; and (3) the lack of clear standards for the issuance of presidential disaster declarations. In essence, the Task Force identifies a lack of consistent policies on when federal assistance is really needed, how much the federal share should be, and whether disaster assistance should be treated as "off-budget" and therefore exempt from deficit-reduction goals.

Three broad approaches to modifying federal disaster assistance were proposed by the Senate Task Force: (1) establish or strengthen criteria for declaring a disaster; (2) create incentives for mitigation; and (3) place increased reliance on insurance. It provided no details as to how any of these should be achieved, and in particular it does not support the concept of all-hazard insurance as recommended by the House Task Force.

The Senate Task Force touched on a key issue in federal disaster policy ignored in the House report, namely the paradox that the federal government is politically constrained from attaching many strings to the disaster assistance it provides.

> . . . The federal role has increased since the mid-20th century to the point that the federal government stimulates and guides states' and localities' planning efforts, appears to provide much if not most of the response and recovery funding, coordinates all response efforts . . . and funds mitigation efforts. . . .
>
> Despite the federal government's apparently dominant functions, federal policies are still intended to supplement the efforts of the nonfederal organizations. *The federal government has relatively little authority over critical components of loss control and emergency management, such as . . . the establishment and enforcement of land-use decisions and building codes.*[82]

In other words, there is a discontinuity between the rising level of federal *fiscal* liability for disaster costs and the diminishing extent of federal *political* influence over development decisions that affect hazard vulnerability. The Senate report made no dramatic proposals to address this paradox, however. (The role of property rights and the "takings issue" in hazard policy are the subjects of Chapters 4 and 5, respectively.)

The NFIP Reform Act of 1994

Congress in late 1994 adopted the National Flood Insurance Reform Act of 1994 (NIFRA)[83] after years of controversy over proposed amendments to limit NFIP coverage in critical erosion hazard zones along the nation's coasts. The proposal was based on a report of the National Academy of Sciences[84] commissioned by FEMA and was adopted by the House of Representatives in 1991.[85] Subsequently, coastal property rights interests succeeded in diluting the amendment in the Senate. As finally adopted, Section 577 of the act charged FEMA with conducting a new study of the economic impacts of mapping erosion hazard areas (not completed as of mid-1998). (See Chapter 6 for more detailed discussion of this controversy.)

While Congress thus rejected any limits on coastal property rights, it endorsed hazard mitigation in other forms. Specifically, NFIRA:

- Strengthened lender compliance with requirements that loans secured by floodprone property are covered by flood insurance policies.
- Provided NFIP coverage for "increased cost of compliance" in bringing floodprone structures up to code.
- Established a Flood Mitigation Assistance Program, financed out of flood insurance premiums to cover 75 percent of the cost of state or local mitigation projects, including: (a) demolition or relocation of structures on eroding shores; (b) elevation, relocation, demolition, or floodproofing of structures in floodplains; (c) acquisition of floodprone properties; (d) minor physical mitigation projects; and (e) beach nourishment activities.[86]
- Authorized a "Community Rating System" (CRS) allowing flood insurance rates to be reduced in communities that exceed minimum NFIP floodplain management requirements.[87]

The 1994 Unified National Program

The original Unified National Program for Managing Flood Losses, transmitted to Congress by President Johnson in 1966, has undergone four revisions, most recently in 1994. The "Unified National Program" is not really a program, as such, and is not self-executing. It is basically a policy statement that summarizes existing roles and responsibilities of the federal, state, and local levels of gov-

areas, as distinct from less hazardous portions of floodplains ("floodway fringe areas"), was a significant loophole in the rhetoric of floodplain management.

In 1994, the loophole became a chasm. The Unified National Program for Floodplain Management retained only its name from earlier iterations. As completely redrafted under the Clinton administration, all pretense of land use regulation as a prerequisite to federal assistance was abandoned. Goal 3 proposes to "Develop and implement a process to encourage *positive attitudes* toward floodplain management" (emphasis added). Supporting "objectives" emphasize "public understanding," task sharing, and incentives. (The latter were reflected in the Community Rating System codified by the 1994 National Flood Insurance Reform Act, as mentioned earlier.) National policy on land use in flood hazard areas had thus shifted from mandatory local regulation pursuant to minimum federal standards to a cultivation of "positive attitudes" and "public understanding." One is tempted to draw an analogy to comparable changes in child-rearing philosophy between the 1960s and the 1990s!

The FEMA Inspector General Report

FEMA has seldom assessed its own programs. This has meant that the agency does not know which of its programs or activities are problematic and which ones are faring well. In 1994, the FEMA Inspector General conducted one of the relatively few self-evaluations that FEMA has initiated, at the request of Richard T. Moore, Associate Administrator for Mitigation.[89] The report identified three channels of FEMA funding for state and local mitigation activities: (1) funds provided under Comprehensive Cooperative Agreements (CCA's) amounting to a total of only $54 million for mitigation between 1988 and 1994; (2) the Hazard Mitigation Grant Program (HMGP) ($334 million available but only $134 million obligated during the same period); and (3) Individual Assistance (IA) and Public Assistance (PA) to a limited extent.

The report noted that about 80 percent of FEMA-sponsored mitigation projects are structural projects related to floods, mostly funded under the HMGP, such as enlarging culvert and storm water drainage systems, elevating highways and bridges, and improving dams and levees. Since other federal agencies also support these types of projects, the report suggested that FEMA should apply its efforts to "relocations, mitigation planning, and public information." It praised the recent effort to buy out thousands of chronically floodprone homes after the Midwest Floods of 1993. (p. 37) But it was critical about four important matters.

1. The Hazard Mitigation Grant Program created under the Stafford Act in 1988 is *unduly cumbersome and time-consuming.* An average of two years is required to process state requests. Only about $145 million out of $334 million available under Stafford Act since 1988 for HMGs was obligated by July 1994. Environmental assessments further slow down the review process.

2. *Inadequacy of state hazard mitigation plans.* The report said the mitigation plans prepared at the state level are not useful, and "plan quality varies because mitigation planning for most of the states we visited was done to qualify for FEMA mitigation grants, rather than to proactively develop a plan for measuring a state's progress in mitigation." (pp. 13–14)

3. *Interagency Hazard Mitigation Teams* are convened hastily and with only a few days available. Team members, drawn from various agencies and levels of government, "often bring their own agendas, and with no more than two days, often come up with unrealistic recommendations. Furthermore, there is no follow-up." (pp. 14–15)

4. *Measuring mitigation effectiveness.* The report states that FEMA has not developed a program to measure the effectiveness of mitigation. It cites the reasons given by mitigation officials at FEMA and in the states; namely, managers do not have the quantitative tools they need to measure mitigation, and they do not have the resources to conduct after-action studies. (p. 35)

More broadly, the FEMA IG Report echoed the NEHRP Advisory Committee in deploring the difficulty of eliciting hazard mitigation at the state and local levels, aside from what the federal government pays them to do.

> A major impediment to the success of mitigation has been state and local governments' unwillingness to commit resources. . . . State officials, particularly elected officials, do not consider it to be a high priority. Funding is more likely to go to projects with a short-term payback. Also, mitigation policies frequently have a negative economic impact because they discourage development in hazard-prone areas. *It will be difficult to increase mitigation efforts unless that attitude is changed* (emphasis added). (p. 18)

National Mitigation Strategy

The mission of FEMA's Mitigation Directorate is to "develop, coordinate, support and implement policies, plans and programs to eliminate or reduce the long-term risk to human life and property from natural and technological hazards; and to support the Director in making mitigation the cornerstone of emergency management."[90] During late 1994 and early 1995, the directorate conducted a series of 11 regional forums to solicit views of state and local emergency managers, public officials, and the general public. These led to the publication of the "National Mitigation Strategy,"[91] which was unveiled at FEMA's first National Mitigation Conference in December 1995.

The National Mitigation Strategy declares mitigation to be the "cornerstone"

of emergency management with two primary goals: (1) increasing public awareness and (2) reducing loss of life and injuries.[92] The strategy strongly reflects the
"partnership" ethic of the federal government in the 1990s (also reflected in the
Galloway Report and the 1994 Unified National Program).

> The foundation of the strategy involves strengthening and creat-
> ing partnerships where none exist in order to empower all Amer-
> icans to fulfill their responsibility for building safer communities.
> These partnerships are needed to address the five major elements
> of the Strategy:
>
> - Hazard identification and risk assessment
> - Applied research and technology transfer
> - Public awareness, training, and education
> - Incentives and resources
> - Leadership and coordination

Significantly, the strategy downplays a federal role in setting standards or
otherwise preempting state or local discretion in hazard mitigation. There is no
mention of the existing mandate for FEMA to set minimum floodplain management standards that communities must adopt and enforce in order to participate in the National Flood Insurance Program. In general, the strategy is
expressed in broad principles without reference to specific programs and initiatives currently in progress.

Initially, FEMA planned to convene biennial strategy meetings on mitigation. But a second national meeting never occurred. Various single topic,
regional meetings were convened. Rather quickly the Mitigation Strategy talk
faded and a new initiative, Project Impact, received priority attention within
FEMA.

Project Impact

The latest in the Clinton era initiatives in hazard mitigation is "Project Impact,"
also known as a program to promote "disaster resistant communities." The
approach is to target FEMA's mitigation efforts upon selected communities,
including a few counties, which have a history of natural disasters and an
expressed intent to reduce future vulnerability. Project Impact envisions the
fostering of partnerships between various stakeholders—federal, state, local,
private—within such communities to implement hazard mitigation programs and measures. Among the first round of Project Impact communities
are Oakland, California; Wrightsville Beach, North Carolina; Deerfield
Beach, Florida; and Allegheny County, Maryland. As of November 1998, the
program was funded at about $150 million per year and had over 100 partici-
pant communities.

CONCLUSION

This chapter has traced the zigs and zags of federal policies, programs, and pronouncements concerning hazard mitigation. The story began in the 1960s with a "brave new world" of nonstructural approaches to disaster loss reduction. From its inception with the National Flood Insurance Program, mitigation soon was incorporated into the disaster assistance program in 1974 and the NEHRP program in 1977. The Stafford Act in 1988 amended the disaster assistance program creating the Hazard Mitigation Grant Program (Section 404). After the Midwest Floods of 1993, Section 404 was expanded to enable FEMA to buy out many thousand flooded dwelling units and to raise the federal share available for hazard mitigation in declared disaster areas. The 1994 NFIP Reform Act established the Mitigation Assistance Grant Program to conduct flood-related mitigation activities regardless of a declaration. Meanwhile since 1993, the new Mitigation Directorate has been occupied with developing the "National Mitigation Strategy" and Project Impact.

There are many indicators of a sea of change over the past three decades in federal policy regarding disasters, from a purely reactive mode to a more proactive perspective, in which the federal government assists states, local communities, and the private sector to anticipate and avoid disaster losses, not simply to recover from them.

Yet certain fundamental issues raised in the various critiques discussed earlier remain unresolved, and perhaps even unaddressed. Foremost among these is the basic question as to the appropriate division of responsibility—fiscal and political—between the federal government on the one hand and the rest of American society on the other. As observed in the National Performance Review, the Galloway Report, and the House and Senate Task Force reports, the prospect of generous federal assistance deters states and local governments from accepting a fair share of the burdens of preparing for and recovering from foreseeable disasters. As stated in the National Performance Review:

> The system as a whole encourages state and local elected officials to ask for maximum federal disaster assistance. Requests have increased approximately 50 percent over the last 10 years. Even minor emergencies have been awarded full compensation. The federal government expects to pay 17 states and the District of Columbia a total of $126 million for the costs of snow removal . . . after the March 1993 storms, even though snow is a regular, recurring event in many of the affected states.[93]

Closely related is the reluctance of states and local governments to adopt and enforce rigorous land use and building regulations to control new construction

vide for their own safety. Communities that fail to comply with floodplain management standards established by FEMA may be suspended from the program and lose their eligibility for flood insurance. Few have been suspended. State or local recipients of disaster assistance that fail to assess and take action to mitigate their natural hazards under Section 409 of the Stafford Act may lose their eligibility for further disaster assistance. But the Section 409 requirements have largely received lip service.

One would hope that such broad policy documents as the "Unified National Program for Floodplain Management" and the "National Mitigation Strategy" would convey a sense of urgency regarding the sharing of burdens of disasters. While stressing the need for partnership, the original principle that federal disaster assistance is merely supplementary to state, local, and private resources is muted. So too is the message that part of the nonfederal effort must comprise accepting the burden (political and economic) of limiting new development and redevelopment in highly hazardous locations such as eroding coasts and narrow canyons subject to landslide or mudslide.

The strategy declares that "all mitigation is local." But unrefuted is the implicit message that *all mitigation is federally funded.* Individual and community responsibility, a primary theme of most of the policy critiques reviewed earlier, must also be the core of the National Mitigation Strategy and actions ensuing therefrom.

But the perception of federal benevolence discourages responsible hazard mitigation among nonfederal interests, thus contributing to the potential for greater losses in future disasters. Shirking responsibility for hazard mitigation among states and local governments may take two forms: (1) unwillingness to expend their own funds for disaster planning and hazard mitigation and (2) avoidance of the political and fiscal burdens of regulating land use in areas subject to natural hazards.

FEMA's goal to place mitigation at the heart of federal disaster policy is thus continually undermined by (1) other federal policies that inadvertently promote building and rebuilding in hazardous areas; (2) federal readiness to assume the lion's share of state and local disaster costs, even for repetitive disasters; and (3) hesitation at all levels of government to enforce effective land use controls in areas of known hazard. The last of these issues is the subject of the next two chapters. Then, a series of case studies will examine how all of these issues have influenced the incorporation of mitigation, or lack thereof, in recovery from three selected disasters.

NOTES

1. Rutherford H. Platt (1998). "Planning and Land Use Adjustments in Historical Perspective," in Raymond J. Burby, ed., *Cooperating with Nature: Confronting Natural Hazards with Land Use Planning for Sustainable Communities.* Washington, D.C.: Joseph Henry Press, 29–56.

2. David Alexander (1993). *Natural Disasters.* New York: Chapman and Hall.

3. Peter May (1985). *Recovering from Catastrophes: Federal Disaster Relief Policy and Politics.* Westport, Conn.: Greenwood Press, Ch. 1.

4. Interagency Flood Management Review Committee (1994). *Sharing the Challenge: Floodplain Management into the 21st Century—A Blueprint for Change.* (Galloway Report). Washington, D.C.: U.S. Government Printing Office, Gl 6.

5. Ian Burton, Robert W. Kates, and Rodney Snead (1969). *The Human Ecology of Coastal Flood Hazard in Megalopolis.* Research Paper No. 115. Chicago: University of Chicago Department of Geography.

6. National Academy of Sciences (1994). "Facing the Challenge;" the U.S. National Report to the IDNDR World Conference on Natural Disaster Reduction, Yokohama, Japan, May 23–27, 1994, 3.

7. Raymond J. Burby (1994). "Floodplain Planning and Management: Research Needed for the 21st Century," in *Water Resources Update,* Issue no. 97, Autumn, 44–47.

8. Peter J. May and others (1996). *Environmental Management and Governance: Intergovernmental Approaches to Hazards and Sustainability.* London and New York: Routledge. The transition referred to is evident in the contrast between the favorable discussion of regulation by May in 1985, note 3, and his 1996 book that refers to the Florida coastal regulatory program as "coercion and prescription."

9. In 1993, Congress raised the share of Stafford Act public assistance funding that can be applied to mitigation from 10 percent to 15 percent. Some states, such as North Carolina and California, which have qualified for hundreds of millions of dollars for mitigation, have found it difficult to actually spend the money on eligible mitigation projects.

10. Rutherford H. Platt and others (1992). *Coastal Erosion: Has Retreat Sounded?* Program on Environment and Behavior Monograph No. 53. Boulder: University of Colorado Institute of Behavioral Science.

11. Rutherford H. Platt (1986). "Floods and Man: A Geographer's Agenda," in Robert W. Kates and Ian Burton, eds., *Geography, Resources, and Environment,* Vol. II. Chicago: University of Chicago Press, 48. White's doctoral dissertation: *Human Adjustment to Floods* (University of Chicago Department of Geography Research Paper No. 29) was the blueprint for this research program and its policy outgrowths. Other influential studies on floods during this period included Luna Leopold and Thomas Maddock, Jr. (1954), *The Flood Control Controversy.* New York: Ronald Press; William G. Hoyt and Walter B. Langbein (1955), *Floods.* Princeton: Princeton University Press; and the work of resource economists Otto Eckstein, John V. Krutilla, Arthur Maass, and others based at Harvard University.

12. Francis C. Murphy (1958). *Regulating Flood Plain Development.* Research Paper No. 56. Chicago: University of Chicago Department of Geography.

13. PL 89–339, Sec. 5.

14. U.S. Congress, Senate Committee on Banking and Currency (1966). *Insurance and Other Programs for Financial Assistance to Flood Victims.* Washington, D.C.: USGPO, 105.

15. U.S. Congress (1966). *A Unified National Program for Managing Flood Losses* (House Document 465, 89th Cong., 2d Sess.). Washington, D.C.: USGPO.

16. Ibid., 1.

17. Ibid., 1 and 16.

18. Ibid., 18.

19. Ibid.

20. Ibid., 34.

21. Ibid., 17–18. The task force suggested that a national flood insurance program be tested on a pilot basis in specific localities to see what its effects on flood losses would be before being applied to the nation as a whole.

22. E.O. 11296 (1966).

23. Robert C. Wood (1967). Testimony re National Flood Insurance Act of 1967 before the House Committee on Banking and Currency (90th Cong., 1st Sess.). Washington, D.C.: USGPO.

24. President of the United States (1973). *New Approaches to Federal Disaster Preparedness and Assistance.* House Doc. 93-100 (93rd Cong., 1st Sess.). Washington, D.C.: USGPO

25. PL 93-234, sec. 2(b).

26. 42 USCA sec. 5176.

27. PL 93-251.

28. Rutherford H. Platt (1979). *Options for Nonstructural Flood Mitigation.* Washington, D.C.: U.S. Water Resources Council.

29. Office of Management and Budget Directive of July 10, 1980, as reflected in Interagency Agreement for Nonstructural Damage Reduction Measures dated December 15, 1980.

30. PL-95-124, as amended; 42 USCA secs. 7701 et seq.

31. 42 USCA sec. 7702(4).

32. From "Remarks Announcing Reorganization Plan No. 3 of 1978," June 19, 1978; and "Message to the Congress Transmitting Reorganization Plan No. 3 of 1978, in Public Papers of the Presidents, 1978, Vol. I, pp. 1128–1131, as cited in the NAPA report, "Coping with Catastrophe" (1993), 15.

33. Sheaffer and Roland, Inc. 1981. *Barrier Island Development Near Four National Seashores.* Washington, D.C.: mimeo; H.C. Miller, "The Barrier Islands: A Gamble with Time and Nature," *Environment* (November 1981).

34. U.S. General Accounting Office, *National Flood Insurance: Marginal Impact on Flood Plain Development* (GAO/CED-82-105). Washington, D.C.: GAO.

35. Concerned Coastal Geologists, "Saving the American Beach: A Position Paper." Skidaway Institute, Georgia: mimeo. This group, which was convened by Duke University coastal geologist Orrin H. Pilkey, Jr., issued a second position paper in 1985.

36. PL 97-35, sec. 341.

37. PL 97-348.

38. PL 97-348, sec. 2(a)(4).

39. PL 97-348, sec. 5.

40. PL 101-591.

41. R.H. Platt, B. Burbank, and S.J. Pelczarski, eds. (1987). *Cities on the Beach: Management of Developed Coastal Barriers.* Research Paper No. 224. Chicago: University of Chicago Department of Geography.

42. Hurricane Andrew in 1992 inflicted some $30 billion in public and private costs even though it missed Miami, Tampa Bay, and New Orleans. If it had directly hit any of those metropolitan areas, the costs would have been considerably higher.

43. Federal Interagency Task Force on Floodplain Management. *Floodplain Management*

in the United States: An Assessment Report Vol. I—Summary Report (Boulder: University of Colorado Natural Hazards Information Center, 1992) and Vol. II—Full Report (U.S. Government Printing Office, 1992).

44. Ibid., Vol. 1, 4.

45. National Review Committee, "Action Agenda for Managing the Nation's Floodplains," reprinted as Appendix F of *Floodplain Management in the United States,* Vol. II.

46. Ibid., F-4.

47. PL 100-707, as codified at 42 USCA secs. 5121 et seq.

48. PL 100-707, sec. 409.

49. PL 100-707, sec. 404.

50. Edward J. Kaiser and R. Matthew Goebel (1996). "Analysis of Content and Quality of State Hazard Mitigation Plans under Section 409 of the Stafford Act." Natural Hazard Working Paper No. 3. Chapel Hill: University of North Carolina Center for Urban and Regional Studies.

51. Ibid., 1–2.

52. GAO (1991). *Federal, State, and Local Response to Natural Disasters Need Improvement: Report to the Congress.* GAO/RCED-91-43. Washington, D.C.: GAO.

53. Ibid., 20.

54. National Earthquake Hazards Reduction Program Reauthorization Act, PL 101-614.

55. 42 USCA sec. 7701(5).

56. 42 USCA, sec. 7701 (6).

57. 42 USCA, sec. 7705b(b).

58. NEHRP Recommended Provisions for the Development of Seismic Regulations for New Buildings Part 1: Provisions (1991) and Part 2: Commentary (1992). FEMA documents no. 222 and 223; prepared by the Building Seismic Safety Council, Washington, D.C.

59. Improvements in building practices regarding seismic safety were proposed in 1985 by an expert panel convened to help provide guidance to FEMA, published as "An Action Plan for Reducing Earthquake Hazards of New Buildings" (FEMA-90, 1985) and *Proceedings: Workshop on Reducing Seismic Hazards of Existing Buildings* (FEMA-91, 1985). The Action Plan identified strategies, techniques, research needs, and resources required to initiate and sustain hazard abatement programs for seismically hazardous nonfederal buildings in the United States. The proceedings included issue papers and a comprehensive review of the draft action plan. The work that was initiated and paid for by FEMA in order to create seismic provisions for new buildings and gain consensus of the building practices professionals use has not been well documented. In this case FEMA has spent several million dollars to develop the provisions for new buildings, but that money has generated a great deal of voluntary work on the part of building design professions— it is estimated that a multiplier of three or four times the investments is likely according to Ugo Morelli, program manager at FEMA responsible for the development of seismic provisions.

60. Action Plan, note 59, 20.

61. Ibid., 33. With respect to NEHRP, two executive orders were promulgated: E.O. 12699 (1990) and E.O. 12941 (1994). The first concerns seismic-resistant standards for new buildings that were owned, leased, or financially supported by the federal govern-

PART II

PROPERTY RIGHTS AND THE TAKINGS ISSUE

For as long as the federal government has responded to natural disasters in the United States, policy makers have chanted the mantra of "land use regulation" as a means to restrain unwise development or redevelopment in areas known to be seriously hazardous. Land use regulations, when effectively used, may limit or entirely prohibit development of sites with extreme vulnerability to flood, earthquake, landslide, or other recurrent natural hazard. This transfers much of the burden of avoiding disaster costs to the owner of the affected property whose land is rendered less valuable if natural limitations are thus explicitly recognized. If the development is canceled or located elsewhere, part of the cost is also transferred to the local taxing jurisdiction that ironically plays the conflicting roles of administrator of local land use regulations and partner through taxation in the development enterprise.

In common with land use regulations of other types (e.g., single use zones, wetlands, habitat protection, historic preservation), natural hazard regulations may incur the wrath of affected private owners and their allies in local government. In the 1990s, the property rights movement has challenged public land use regulations of many types through political action, litigation, public outreach, and, in extreme cases, intimidation and violence. The more responsible participants in the property rights movement have effectively imitated the environmental movement in resisting what are thought to be unwarranted encroachments by government upon the prerogatives of private land ownership. The standard objection raised is that regulation "takes" private land without compensation in violation of the Fifth Amendment to the U.S. Constitution.

Chapter 4 provides an overview of the property rights movement, circa 1995, as a prelude to a detailed analysis of the "takings issue" in relation to natural hazards, which is the subject of Chapter 5.

Property Rights Organizations: Backlash Against Regulation

Jessica Spelke Jansujwicz

INTRODUCTION

Chapter 3 recounted the rise and fall of noncompensatory land use regulation as a tool for hazard mitigation. Compulsory local regulation of new construction in high hazard areas was an essential quid pro quo of the National Flood Insurance Program, as established in 1968. This expectation was reinforced by the Flood Disaster Protection Act of 1973 and the Federal Disaster Relief Act of 1974. Today, Section 409 of the Stafford Act requires recipients of federal disaster assistance to evaluate natural hazards in the areas affected by a disaster and take action "to mitigate such hazards, including safe land use and construction practices."[1]

While "safe construction practices," most notably minimum elevation requirements and strengthening against wind damage, are increasingly required in coastal hazard areas, efforts to limit further encroachment through land use zoning, wetland restrictions, or other noncompensatory limits on the location of development have been soft-pedaled by governmental agencies in the 1990s. Instead, as argued in Chapter 3, mitigation has increasingly come to mean projects funded in large part by the federal government rather than outright limitation on building or rebuilding in areas of significant natural risk. Many (but not all) states and local governments share this hesitation to exert noncompensatory authority over land use that may lower private property values and community tax revenue. This apparently reflects a heightened fear that noncompensatory regulations may be held to be a "taking" of private property values in violation of the Fifth Amendment of the U.S. Constitution, which states in part: "Nor shall private property be taken for public use without just compensation." According to Peter Douglas, executive director of the California Coastal Commission: "Environmentalists and those of us in government responsible for environmental stewardship have been outgunned, outflanked, and

111

generally overwhelmed by the tactics and achievements of the private land rights coalitions."[2]

Much of the responsibility for placing land use regulation on the defensive may be attributed to an aroused property rights movement that erupted across the United States in the early 1990s. Dozens of new grass-roots groups were organized in opposition to the power of government to regulate private property for environmental or other purposes without compensation. Increasingly, property rights organizations (PROs) have gained political influence and visibility at the federal, state, and local levels. To an unknown degree, they constitute a potential obstacle to natural hazard regulations that limit the use of land in coastal and riverine floodplains, in areas subject to landslide, mudslide, or wildfire, or in areas of seismic hazard.

Increased visibility of the property rights movement can be attributed to several factors: (1) conservative Republican control of Congress after the November 1994 elections, (2) growing impatience of many American property owners and businesses concerning the economic costs of government regulations, (3) an apparent arrogance displayed by certain environmental agencies and bureaucrats, and (4) the short-lived political meteor known as the "Contract With America." Takings compensation was a provision of the "Contract" that would require the federal government to compensate owners for lower property values resulting from federal regulations that prevent them from developing land considered "environmentally sensitive."

The property rights movement is closely allied with the "wise use" movement, which challenges restrictions on private uses of the resources of federal lands in the West, including forest products, grazing, water, minerals, wildlife, and commercial recreation development. This chapter will focus on the property rights movement, although some of the organizations reviewed also may identify themselves with the "wise use" movement.[3]

The constitutional issues surrounding the "taking issue" and natural hazard regulation are considered in more detail in Chapter 5. This chapter will take a brief look at some of the more prominent property rights organizations and their beliefs. An inventory conducted by Greenpeace[4] identified a diversity of PROs located across the United States, with several concentrated in Washington, D.C. While these PROs typically take the form of nonprofit corporations, similarities generally end there. Staff, funding, activities, and political positions vary considerably among PROs surveyed. For all their influence, most PROs are small and spartan. Staff size among those examined ranged from only three employees to approximately 27. Office spaces range from one or two cramped rooms to a suite of spacious offices. Funding for staff and office space originates from a variety of sources including membership dues, foundation grants, corporate partners, publication royalties, and/or services rendered.

Each PRO molds its agenda, programs, and expertise to fit the needs of a specific constituency. For instance, several PROs use property rights as their central organizing principle. Others reach larger audiences by addressing wider-ranging issues such as regulatory takings, employment discrimination, and management of municipal solid waste. Some PROs target specific legislative initiatives (e.g., Section 404 of the Clean Water Act and the Endangered Species Act). Others focus on a particular level of regulation (e.g., federal, state, local).

All PROs seek to influence public opinion and political action in various ways. Their activities include at least some of the following: (1) publications, including newsletters, pamphlets, articles, and books; (2) informal lectures to public officials and the academic community; (3) formal testimony; (4) lobbying; and (5) litigation. PROs differ widely in the particular range of activities utilized. While some PROs specialize in one particular area, several use a combination of approaches to achieve their ultimate goal. The following PROs were selected for more detailed analysis.

Washington-based Property Rights Organizations

- Competitive Enterprise Institute
- Defenders of Property Rights
- National Legal Center for the Public Interest
- The National Wetlands Coalition

Property Rights Organizations Based Elsewhere

- Pacific Legal Foundation
- Political Economy Research Center
- Fire Island Association
- Save Our Seashores

These organizations represent a wide diversity of functions, mode of organization, budget, and scope of influence. Each was studied through face-to-face or telephone interviews, mail questionnaires, and review of their written documents. Small advocacy organizations such as PROs (and environmental organizations for that matter) are subject to rapid change in terms of funding, leadership, focus, and office location. The following profiles portray these organizations as of the mid-1990s.

It should be noted that the focus of most of these organizations is not primarily on natural hazard regulations. But the efforts of these PROs and their allies to promote a narrower permissible scope for governmental "takings" casts a broad cloud over land use regulation affecting natural hazards as well as their

otherwise staying out of our lives." Ultimately, DPR seeks to "halt the growth of government and restore power to the people."[16]

DPR collects a membership fee from approximately 2000 members. Additional funds are raised from foundation donations and legal fees (nonprofit status permits DPR to charge clients a base price for services). DPR has a small full-time staff, supplemented during the summer months by legal interns.

DPR uses a three-pronged approach to meet its objectives:

1. *Litigation.* DPR seeks "equal protection" for all landowners by representing small property owners who typically cannot pay for lawsuits. DPR identifies and accepts cases that may further their policy objectives.
2. *Education.* DPR publishes and distributes a newsletter to 3000–5000 individuals. The media (e.g., radio, magazine and newspaper articles) plays an important role in promoting DPR's grass-roots image. Several recent articles recount anecdotes concerning alleged abuse of property rights by government agencies. Others draw comparisons between property rights and civil rights.[17]
3. *Legislation.* While DPR does not consider itself a lobbying organization, it pursues program objectives through legislative analysis.

DPR strongly opposes the wetland regulatory program under Section 404 of the Clean Water Act. Roger J. Marzulla argues that the wetlands permitting program "purports to regulate virtually all land-disturbing activities occurring on 100 million acres or more of privately owned property." In instances where land-disturbing activity is "confined to the boundary line of the property itself," Marzulla argues that ". . . there is no rational—not even common law nuisance—for the environmental regulatory and enforcement bureaucracy to usurp the owners' property rights." Furthermore, he adds ". . . the regulation of a wetland does not substantially advance a governmental interest in protecting its citizens against pollution or obstruction of navigable waters."[18]

DPR also believes "The government's implementation of the Endangered Species Act (ESA) has served to slow economic growth and take private land without just compensation."[19] The ESA, riddled with "liberal interpretations" and "voluminous listings,"[20] is a "classic example of giving no regard to property rights."[21] Liberal interpretations of the "takings" prohibition are particularly offensive to Marzulla who accuses the federal Fish and Wildlife Service of using this clause "to assert control over a wide range of private activity on private lands."[22] Two recommendations Marzulla makes for federal agencies are (1) to concentrate on protecting species rather than subspecies and (2) to provide a practical opportunity to resolve conflicts between the rights of property owners and the listing process.[23]

Regarding coastal development, DPR believes that a "one-time" hurricane is not enough cause to restrict building in a coastal area. Instead, the "right to rebuild" question should be evaluated on a case-by-case basis. Landowners should have the right to rebuild in a coastal hazard zone at their own risk. According to DPR, landowners are not "double-dipping" when compensated by the federal government for coastal disaster losses. Rather, "amenities are usually paid for by the property owners." In other words, "bridges do not magically appear—they are paid for by taxes imposed upon productive members of society such as property owners."[24]

In an attempt "to filter out some of the distortions and misconceptions about property rights legislation," Nancie Marzulla testified before the Senate Judiciary Committee on April 6, 1995, that "Protection of private property rights need not be the enemy of achieving important social objectives." Marzulla argues that "individuals guided by free market incentives are often better stewards of the environment than the heavy hand of government." Therefore, "private property owners protect their property from environmental harm with greater vigor than the government."[25]

Roger Marzulla, as assistant attorney general for the Land and Natural Resources Division of the Justice Department in the 1980s, drafted President Reagan's Executive Order 12630.[26] This order imposed on federal agencies a duty to protect property rights in the course of administering its regulatory programs. Essentially, the order requires that a takings impact analysis be prepared by federal agencies "before undertaking any proposed action regulating private property use for the protection of public health or safety."[27]

Nancie Marzulla supports similar models for state property rights bills. She advocates two approaches to protect property rights through state legislation: (1) "Look Before You Leap" bills and (2) takings compensation bills. "Look Before You Leap" legislation is patterned after federal Executive Order 12630 and requires state agencies and/or local governments to assess the takings implications of state regulations before they are adopted. Takings compensation bills define a taking by creating a "trigger-point" at which a regulation is presumed to have become a taking. These bills entitle a property owner to automatic compensation on proof that a government regulation reduced the value of specified property by a certain percent. DPR has drafted model legislation for states and Congress that would define a taking as a diminution in value of 50 percent or more of preregulatory value.[28]

National Legal Center for the Public Interest

The National Legal Center for the Public Interest (NLCPI) was established in 1975 by a group of conservative businessmen.[29] Its mission is "to foster knowledge about law and the administration of justice in a society committed to the rights of individuals, free enterprise, private ownership of property, balanced use of private and public resources, limited government, and a fair and efficient judiciary."[30] NLCPI pursues a wide-ranging agenda including:

- regulatory takings
- alternative dispute resolution
- employment discrimination issues
- natural resources damages
- sentencing guidelines for environmental crimes
- "uses and abuses" of science in litigation
- agricultural and environmental law
- emerging antitrust litigation

NLCPI does not seek or accept state or federal funding. Instead, NLCPI is totally dependent on voluntary contributions from sources such as AT&T, Ford Motor Company, Exxon, the Hearst Foundation, and the John M. Olin Foundation.[31] As of 1995, NLCPI staff consisted of only three officers—a president and two vice presidents. Because NLCPI staff is limited, both vice presidents perform several roles. One serves as general counsel and assistant treasurer while the other covers administration.

NLCPI does not litigate and is not a "grass-roots" organization. Rather, the organization takes a "nonlobbying approach" fulfilling its missions through the publication of scholarly monographs, a law review, manuscripts, and a newsletter. NLCPI also organizes forums, government briefings, and press briefings to bring together senior federal governmental and judicial officials and organization members.

NLCPI primarily targets Section 404 wetland regulations, Superfund, RCRA, and the Endangered Species Act.[32] Regarding coastal development, NLCPI General Counsel Roger Clegg recognizes gray areas in banning development in unsafe areas. In "gray areas," the best way for the law to work is to determine whether the government is actually using the property on which the regulations were imposed. In the case of *Lucas v. South Carolina Coastal Council*[33] (discussed in Chapter 5), for example, Clegg argued that building restrictions were imposed so Lucas's property could effectively be used as a buffer zone, which is a legitimate public use, "but one for which compensation must be paid by the owner."

The National Wetlands Coalition

The National Wetlands Coalition (NWC) was established in 1989 by a group of corporate interests, including utility, mining, and real estate, who retained the Washington law firm of Van Ness, Feldman and Curtis to lobby Congress on wetland issues.[34] Its mission is to work with Congress and the administration for legislative reform of the federal wetlands permitting program, authorized by Section 404 of the Clean Water Act. The coalition supports the management of wetlands "in a manner that effectively and sensibly accommodates the competing, legitimate demands for conservation and use of wetlands resources."[35] According to the coalition, the goal of "no overall net loss of wetlands values" should be pursued by implementation of a variety of regulatory and nonregulatory programs designed to (1) conserve wetlands values and functions to the maximum extent practicable, (2) eliminate or streamline procedures for use of wetlands of marginal resource value, and (3) provide incentives for private wetlands conservation efforts.[36]

The coalition is funded solely by the dues contributions of its approximately 70 member organizations. Coalition membership is diverse and includes representatives of county and parish governments, municipalities, ports, Native American groups, engineering and environmental consulting firms, banks, the development community, and industrial interests. Membership is "composed of those that own land that is characterized as Federal jurisdictional wetlands and those that pursue economic and governmental activities that may be affected by the Federal wetlands regulatory policy."[37] As of 1995, a partial list of coalition members included the American Farm Bureau Federation, American Road and Transportation Builders Association, the Louisiana Land and Exploration Company, National Association of Homebuilders, Natural Gas Supply Association, National Utility Contractors Association, Shell Oil Company, and Union Pacific Resources Company.

The coalition is staffed by the law firm of Van Ness, Feldman and Curtis and is governed by a board of directors. It lobbies on wetland issues and provides many services to support its mission, including (1) developing and supporting legislative and administrative initiatives, (2) summarizing relevant bills that are introduced in Congress, (3) providing research and drafting assistance, (4) answering general wetland inquiries, and (5) organizing and attending periodic meetings with members of Congress and key administration officials in Washington, D.C.

The coalition staff works closely with members of Congress and administration officials. The coalition assisted in the development of and supports the enactment of the Comprehensive Wetlands Conservation and Management Act (H.R. 1330 in both the 102nd and 103rd Congress). The coalition worked with the White House Task Force to develop the president's wetland proposal announced on August 23, 1993. In addition, the coalition testified before several committees of the House of Representatives and the Senate as well as before the Wetlands Task Force of the Bush administration Domestic Policy Council and the Clinton administration's White House Task Force.

The coalition seeks to serve as the "focal point for the regulated community" in the campaign to reform the Section 404 permitting program.[38] The coalition seeks to reform this "very broad" and "rigid" program and thereby "reduce the conflict between our nation's environmental goals and economic and property ownership goals." The coalition believes "this environmental program . . . must evolve away from primary reliance on the 'central command and control' model regulation to a more incentives-based program."[39]

The coalition has not taken a position on coastal development except with reference to wetlands laws that it opposes.

PROPERTY RIGHTS ORGANIZATIONS BASED OUTSIDE WASHINGTON, D.C.

Pacific Legal Foundation

The Pacific Legal Foundation was founded in 1973 as a public interest law firm that studies and publishes reports on a broad array of environmental issues.[40] It is based in Sacramento, California. The foundation's 1994 budget was $3.3 million. Major gifts, generally from corporations and associations, accounted for 26 percent of their contributions. Approximately 32 percent were from smaller individual donations. The foundation supports approximately 12 staff attorneys.

The foundation's primary activity is litigation, along with research and writing. Staff attorneys frequently participate directly in lawsuits as intervenor or as amicus curiae. The foundation also litigates in areas other than environmental law to achieve their goal of limited government and the preservation of individual freedoms. This includes activities supporting tax limitation measures (such as California's Proposition 13), term limits, and activities opposing reverse discrimination, rent control, and coerced speech.

The foundation does not engage in direct legislative activity and, therefore, does not have a "legislative focus" per se. Rather, the foundation monitors topics such as the "takings" issue, federal clean water regulations, wetland protection, wilderness designation, hazardous waste cleanup, and mining on public lands to provide technical advice. It also monitors zoning and Endangered Species Act issues.

In reference to coastal development, staff attorney James Burling feels decisions are best left to the individual rather than to government: "While the government bureaucracy may be well suited to set health and safety standards for building construction, the choice of whether or not to build in the first place is not the role of government unless a clear nuisance or threat to safety is involved." (This public safety/nuisance exception is not often explicitly stated by other PROs.) According to Burling, the government's responsibility is to assure that private property owners are informed of pertinent information that may influence their decisions: "Government action might be more appropriate if it were generally confined to ensuring that property owners, especially prospective owners, are fully informed of the risks." All property is subject to risk, and Burling feels property owners should have the opportunity to understand the risk before making their own choices.

Burling is a proponent of market forces over command and control government regulations. His preference would be "to allow property owners to reasonably develop their properties even if located in regions which may have some potential for damage from natural disasters." He would allow for construction in "areas with some limited potential for fires, hurricanes, flooding, tornadoes, earthquakes, tsunamis, landslides, urban violence, and the like—so long as the development itself will not create a nuisance or threat to life."

Burling generally opposes special subsidies (including disaster relief) for building in potential hazard areas. He further recognizes that "this issue becomes somewhat more ethically complicated for those property owners who have developed in hazard areas in reasonable reliance on government subsidies." In general, Burling is "opposed more to the type of disaster relief that encourages and pays people to rebuild homes and businesses in the same disaster prone area." He is "not opposed to true humanitarian disaster relief that protects public health and safety."

Political Economy Research Center

The Political Economy Research Center (PERC) was founded in 1980.[41] It is based in Bozeman, Montana. PERC's primary goal is "to provide market solutions to environmental problems and to lay the foundations for better policies." PERC advocates free-market environmentalism and private environmental protection, arguing that the environment is better protected in the long run by voluntary action rather than "government coercion." PERC supports policies and programs that conserve wildlife habitat through the promotion of economic incentives.[42]

In 1994, PERC received $611,519 in contributions from foundations, trusts, and endowments. Major contributors include Amoco Foundation, American Forest and Paper Foundation, American Farm Bureau Federation, Pfizer, Inc., M.J. Murdock Charitable Trust, and Independence Mining. PERC has a staff of approximately 10 people and a board of directors.

PERC promotes its agenda through a program of education and advocacy. It seeks to "educate people about alternatives, rather than support specific ones." PERC's staff of academics and associates write prolific amounts of position papers, op-ed pieces, and scholarly articles and books, primarily on issues of federal regulation.

PERC focuses on a number of environmental issues to encourage market-based approaches wherever possible. They seek to "reinvent environmentalism by bringing more reality and common sense to environmental policy." PERC highlights incentives and secure and tradable property rights to encourage cooperation and resource stewardship.[43] These principles are applied to legislative issues including the Clean Water Act, Endangered Species Act (ESA), 1995 Farm Bill, and Superfund.

Regarding coastal development, PERC supports actions to eliminate federal funding for roads, bridges, sewers, water lines, housing, and flood insurance in large areas of coastal land. Eliminating federal subsidies would reduce future disaster costs while not unreasonably limiting private property rights. Because a great deal of development, especially on barrier islands, occurred with the support of federal funds, PERC argues that if federal support is eliminated, much of the incentive to build will be reduced without taking away an individual's property rights. In instances where people want to build and are willing to bear the costs themselves, PERC feels they should be permitted.

PERC feels that eliminating subsidies, including disaster assistance, would change the dynamics of building in many areas. For instance, it would lead people to be more cautious and to deal more responsibly with the potentials of hazards. In other words, it "puts responsibility where it belongs—on the shoulders of the decision makers."[44]

Fire Island Association

The Fire Island Association (FIA) was established in 1955 as the Fire Island Erosion Control Committee.[45] This committee was formed after the Fire Island barrier, which extends over 30 miles along the south shore of Long Island, New York, was seriously damaged by a series of coastal storms. In 1961, a new ad hoc association called the Fire Island Voter Association was formed by leaders of the Erosion Control Committee. In 1962, after a severe northeast storm, the organization became known as the Fire Island Association.[46]

FIA was formed to represent property owners in 17 communities scattered along approximately 12 miles of Fire Island. There are approximately 3700 structures in the 17 communities within the Fire Island National Seashore. Membership in FIA numbers about 1500. Each member pays annual dues of $75. FIA is run by an executive committee headed by President Gerard Stoddard. It operates from an office at Ocean Beach on Fire Island during the summer and from a New York City office during the off-season.

While FIA's "primary concern is for the property rights of the owners of approximately 4000 homes," it recognizes that "the erosion of the barrier island poses an emergent threat to all homeowners and businesses along the north shore of the Great South Bay as well."[47] The Fire Island Association encourages government support of structural measures to protect and restore the integrity of the barrier. At the same time, FIA stresses the importance of maintaining a balance between the interests of private property owners and the interests of the public at large. The "Association believes it is in everyone's long term interest to support positive actions to restore the protective function of the barrier and to support the rights of all owners to continue to use and enjoy their property."[48]

FIA strongly supports government action to assure the integrity of the Fire Island barrier. FIA uses lobbying and public education to solicit support for government action to ensure the integrity of the Fire Island barrier. Staff members, particularly Gerard Stoddard, inform the public through newsletters, articles in the general and academic press, and presentations at civic groups, academic institutions, legislative hearings, and community meetings. FIA has also published policy statements to outline its position on preserving the barrier island system and reconstructing storm-damaged properties.

FIA strongly opposes erosion-based restrictions on new or rebuilt structures under the National Flood Insurance Program (NFIP). Such restrictions were proposed in HR 1236 adopted by the U.S. House of Representatives in 1991 and subsequent bills as recommended by the National Academy of Science Committee on Coastal Erosion Management. FIA, with other coastal property organizations, succeeded in blocking any erosion restrictions or mapping under the NFIP. (See Chapter 6.)

The association supports the reconstruction or relocation of storm damaged houses at the owner's expense. While, FIA makes no reference to sources of financial support for these projects, the Association clearly states that "(p)olicies regarding rebuilding of storm-damaged properties should be enacted so that necessary repairs to damaged structures can be made promptly and in a way that supports efforts to preserve and strengthen the barrier." In the case of stranded houses, "owners should be permitted to retain full control of their property where there is a reasonable prospect that a combination of natural processes and dune building could restore the building site to viability."[49] In instances where rebuilding is not possible, FIA supports voluntary acquisition of flood-damaged properties.[50]

FIA cites the environmental movement as the primary impediment to enhanced government support. Stoddard blames New York State's reluctance to fund shore protection and the resultant decline in spending for shore protection activities to some extent on the "rise of the environmental movement and its strong antipathy toward fixed shoreline erosion control structures."[51] According to Stoddard, environmental groups oppose "hard" shoreline structures such as groins, not because of the harm they may eventually inflict on the environment, but merely because it is a convenient way to stop development in coastal areas.[52] Essentially, "coastal land owners believe most coastal laws and regulations are ultimately aimed at removing private property owners from the coastline."[53] Stoddard supports this claim by arguing that because "groins are designed to slow beach erosion, thus making it less dangerous to build near the shore, . . . people who believe the coastline should be kept free of development of any kind oppose groins for this reason."[54]

While FIA does not directly address property rights legislation, Stoddard portrays property owners as being under siege by the government, not by natural hazards. Stoddard argues that even if policy initiatives, incorporated into local land use control regulation, "aren't actually taking their property, the owners feel certain they will have the effect of seriously devaluing it."[55] Although the "stated objective may be given as the protection of the environment or the protection of public safety . . . the unstated underlying objective is to convert private property into public open space, at the lowest possible cost."[56]

Stoddard argues that the "first obligation of government in coastal management is not to clear the shore of development." Instead, he supports government action focusing on the maintenance of navigational inlets and efforts to ensure actions used to "deter erosion on updrift beaches do not result in the starvation of downdrift beaches of the sand that is vital to the protection of other private and public property."

Save Our Seashores

Another voice on behalf of beach protection, calling itself "Save Our Seashores" (SOS), was established after the winter storms of 1992/1993 devastated the south shore of Long Island.[57] SOS is a nonprofit grass-roots coalition concerned with the economic and ecological consequences of coastal erosion along New York State's Atlantic coastline, including but not limited to Fire Island. The goal of SOS is to "raise the coastal consciousness of all New Yorkers and to preserve the state's precious coastal resources."[58] Through policies and programs, SOS furthers "the concept that a healthy seashore is vital to the economy and the ecology." Executive Director Jane Ekstein states that SOS is not a property rights organization, as such. Rather, SOS is defined as a "broad-based coalition of public officials, civic groups, businesses, and individuals united in the belief that the coastal resources of New York State must be protected and preserved for the common good."

Financial support comes from direct solicitation of the organization's members. SOS also applies for foundation grants and hopes to receive support from corporate sponsors in the near future. SOS staff consists entirely of volunteers. SOS engages in public advocacy targeted at elected officials, towns and villages, civic and homeowner groups, schools, hospitals, businesses, and private citizens. SOS raises visibility for coastal resource protection at assembly hearings and educational outreach programs including informational booths at seashore festivals. In the future, SOS hopes to expand its environmental education programs.

Ekstein stresses the interconnectedness of the environment, the economy of the barrier islands, and the vast estuary system they protect. In light of the close relationship between the economy and the ecology of the area, Ekstein states that " . . . coastal resources must be managed with a proper respect for the right balance, taking into consideration the needs of all those who live at, work at, use or otherwise benefit from the area." Ekstein does not advocate an approach to "armor the coast or [to] try to keep moving land masses in a fixed position."

Instead, SOS seeks to move away from "a patchwork of solutions" to managing the system as a whole. While Ekstein realizes that "truly natural disasters cannot be prevented," she clearly feels that "we can stop helping disasters along." More important, she argues that management plans should "stop focusing on coastal homeowners and look for the broader implications of the choices we have made along our shores."

CONCLUSION

This chapter has provided snapshots of a diverse array of PROs located throughout the United States. While they differ widely in their specific agendas and proposed solutions, basic core beliefs and general purposes of both Washington-based and non-Washington-based PROs reflect several common objectives:

- Limit the power of government
- Promote free enterprise
- Protect private property
- Advocate legislative reform

PROs reflect a common preference for "market-based" solutions in place of governmental "command and control" approaches to managing environmental resources. For instance, several PROs, such as Defenders of Property Rights, in interpreting "takings" law, regard the common law "nuisance" doctrine as sufficient to protect public health, safety, and the environment. The National Wildlife Federation, to the contrary, argues that "common law nuisance offers little, if any, meaningful protection for homeowners and the public," particularly since "proving a common law nuisance can be extremely difficult."[59] For example, in certain states, flooding neighboring homes by filling a wetland may not be considered a nuisance. In the state of Maine, for instance, under the common law of nuisance, a plaintiff must prove that the defendant artificially stored and then discharged water or established a water course (complete with stream bed and banks) onto the plaintiffs' land.

Unfortunately, as discussed in Chapter 5, there is no hard and fast set of rules to determine acceptable levels of nuisance, and the standard of reasonableness or unreasonableness allows courts broad discretion.[60] Although defining classic common law nuisance may be difficult, care must be taken to not define this police power too narrowly and thereby allow injurious uses. As stated by Roger Pilon of the CATO Institute, the nuisance exception should be tied to "a realistic conception of rights."[61] On its face, this statement may convey a relatively simple principle. In reality, however, how does one define "reasonable" when describing a permitted land use? The use of these words by PROs provides little, if any, guidance to resolve the "takings" issue. Instead, it ignores the fact that the reason the government exists in the first place is to define what is for the common good and what is not.

The common law doctrine of nuisance, which dates back to the Middle Ages in England, holds that there is an obligation to use land so as not to harm others. This duty was enforceable by private victims through court action (private nuisance), and also by government through the enactment and enforcement of ordinances (public nuisance). No compensation has ever been required for violation of either type of nuisance. By demanding compensation for landowners who would not ordinarily receive any under nuisance law, PROs are effectively demanding a reward for not causing harm to neighboring landowners and the public at large. Because nuisance law is site-specific, takings claims should continue to be determined by examining site-specific facts on a case-by-case basis. "Trigger points" embodied in property rights legislation supported by PROs provide blanket prescriptions for determining a taking and will do little more than roll back existing environmental laws and make future environmental protection prohibitively expensive. Ultimately, budget constraints imposed by such legislation would force federal agencies to make decisions based on what they can afford, rather than on what the natural resource is environmentally worth. Thus in many instances, PROs may indeed be "wolves in sheep's clothing," protecting industrial or development interests in the guise of supporting the individual property owner.

PROs also ignore important principles in other contexts. For instance, several PROs, particularly DPR, argue that landowners are not "double dipping" when they are compensated by the federal government for certain building restrictions, because amenities (e.g., roads, bridges) are paid for by the taxpayer. In reality, however, this argument ignores the disproportionate burden that is placed on the American taxpayer. Many PROs (e.g., CEI, PERC) do, in fact, recognize this inequality and support elimination of federal subsidies to place the responsibility where it belongs—on the individual landowner. PERC, in particular, reasonably argues that if you remove federal support (subsidies), you will remove much of the incentive to build in coastal hazard areas and other environmentally sensitive areas without taking away an individual's property rights.

"Storytelling" is a popular strategy used by PROs (e.g., CEI and DPR) to attempt to persuade Congress to roll back many environmental protections. The National Wildlife Federation cites several of these "horror stories."[62] (Environmentalists, of course, also may engage in "anecdotal politics.") Although evidence shows that these stories are sometimes exaggerated or even false, PROs effectively use them to portray a helpless victim who was penalized for trying to make his own land useful. By using such scare tactics, PROs highlight public misconceptions and promote what Frank I. Michelman refers to as a "mistakenly oversimplified view of the place of private property rights."[63] In their argument, PROs ignore the fact that the U.S. Constitution does not guarantee that the most profitable use will be allowed, and anyone who purchases property takes the risk that government regulations will diminish its value.

If the property rights movement is based on a growing number of these misconceptions and distortions, what accounts for its apparent success? The property rights movement is not an unpredictable phenomenon, as movements often beget countermovements. Over the last several years, the environmental movement has enjoyed a considerable amount of success. With that success came the strong backlash embodied in the property rights movement. The property rights movement and its membership organizations will continue to exert their influence in a political arena where the American public is increasingly disillusioned by government and anxious about the growing cost of environmental regulation.

This study indicates that property rights organizations at the national, regional, and local scales have become a strong political force to reckon with. Environmental organizations can no longer ignore their effect in the current political arena. This effect, however, may not be entirely negative. In fact, PROs may be a positive influence on the environmental movement. For instance, in the last few years, the apparent success of PROs has forced the environmental movement to reassess its own agenda and reinvigorate its grass roots. By encouraging environmental organizations to seek innovative solutions to today's environmental problems, these disparate groups may be able to reach common ground. After all, protection of private property rights and government regulations should not be mutually exclusive goals. As several PROs acknowledge, promoting private incentives and nonregulatory initiatives (e.g., the U.S. Fish and Wildlife Service's Partners for Wildlife Program) could encourage environmental protection and stewardship. They feel, however, that such nonregulatory programs should replace, not supplement, existing federal regulations.

Finding common ground is essential if natural hazard and environmental goals are to be realized. Based on this study, it is evident that such common ground will be achieved only if exaggerated "storytelling" and misconceptions are eliminated and constructive dialogue is encouraged.

NOTES

1. PL 100-707, Sec. 409; 42 USCA Sec. 5176.

2. As quoted in "'Taking' Back the Initiative," *Daily Journal,* May 12, 1998. According to Douglas, public officials at all levels of government in the West have been intimidated by the threat of "takings" lawsuits and have refrained from controversial regulatory actions (Personal Communication, May 20, 1998).

3. Phil Brick (1995). "Determined Opposition: The Wise Use Movement Challenges Environmentalism." *Environment* 37:17–20; 36–41.

4. Carl Deal (1993). *The Greenpeace Guide to Anti-Environmental Organizations.* Berkeley, Calif.: Odionian Press.

5. Competitive Enterprise Institute (1994). *Annual Report,* 28.

6. Ibid., 7.

7. Ibid., 3.

8. Ibid., 11.

9. Jonathan H. Adler (1995). "The True Nature of Property Rights," *The Wall Street Journal* (February 7), 14.

10. *Babbitt v Sweet Home Chapter of Communities for a Greater Oregon* 115 S.Ct. 2407 (1995).

11. Ike C. Sugg (1995). "Sweet Home, Sweet Justice: Sweet Home vs. Babbitt," in *The Competitive Enterprise Institute Property Rights Reader.* Washington, D.C.: CEI, 32.

12. Ibid., 8.

13. Jonathan H. Adler (1995). "Introduction: Property Rights Fact and Fiction," *The Competitive Enterprise Institute Property Rights Reader,* 5.

14. Jonathan Tolman (1995). "Property Rights and Wrongs," *The Wall Street Journal* (January 16).

15. The following is based on an interview with David Almasi, Media Relations Director, unless otherwise noted.

16. Defenders of Property Rights (1994). *Property Rights Reporter* Fall, 7.

17. Nancie G. Marzulla (1994). "The Civil Rights Issue of the '90s." *The World and I,* (October).

18. Testimony of Roger J. Marzulla before the Senate Environment and Public Works Committee (July 27, 1995).

19. Ibid., 9.

20. Ibid., 11.

21. Ibid., 10.

22. Ibid., 11.

23. Ibid.

24. Defenders of Property Rights. *10 Facts About Property Rights Legislation,* 13.

25. Testimony of Nancie G. Marzulla, President and Chief Legal Counsel of Defenders of Property Rights before the Senate Judiciary Committee (104th Cong, 1st Sess.) (April 6, 1995). Washington, D.C.: U.S. Government Printing Office.

26. "Governmental Action and Interference with Constitutionally Protected Property Rights," Executive Order No. 12,630, 3 *Code of Federal Regulations* 554 (1988).

27. Defenders of Property Rights. *Property Rights Reporter,* (Fall 1994), 5.

28. Hertha L. Lund (1994). "Property Rights Legislation in the States: A Review" PERC Policy Series Issue Number PS-1 (January 1995); Nancie Marzulla (1994). "State Private Property Rights Initiatives as a Response to 'Environmental Takings.'" Chapter 4 in R.M. Clegg, J. DeBow, J. Ellig, and N. Marzulla, eds. *Regulatory Takings: Restoring Private Property Rights.* Washington, D.C.: National Legal Center for the Public Interest.

29. The following is based on an interview and correspondence with Roger Clegg, General Counsel, unless otherwise noted.

30. National Legal Center for the Public Interest (1994). Annual Report, 1.

31. Deal, note 4.

32. Roger Clegg, ed. (1993). *Farmers, Ranchers, and Environmental Law.* Washington, D.C.: NLCPI.

33. *Lucas v. South Carolina Coastal Council* 112 S.Ct. 2886 (1992).

34. The following is based on an interview with Kim Putens, executive director, unless otherwise noted.

35. National Wetlands Coalition (1990). *Statement of Principles* (July 19).

36. Ibid.

37. Ibid.

38. National Wetlands Coalition, note 35.

39. Testimony of Robert G. Szabo, Counsel, the National Wetlands Coalition before the Water Resources and Environment Subcommittee of the House Transportation and Infrastructure Committee, Clean Water Act Reauthorization (March 7, 1995), 4.

40. The following is based on written correspondence with James Burling, staff attorney.

41. The following is based on correspondence with Jane Shaw, senior associate, unless otherwise noted.

42. Terry L. Anderson and Donald R. Leal (1991). "Inside Our Outdoor Policy," in *Free Market Environmentalism.* Boulder, Colo.: Pacific Research Institute for Public Policy Westview Press.

43. Terry Anderson, ed. (1995). *Reinventing Environmentalism in the New Era: Promoting Environmental Quality and Fiscal Responsibility.* PERC Policy Series Issue Number PS-2.

44. Richard L. Stroup (1992). "The Endangered Species Act: A Perverse Way to Protect Biodiversity." PERC Viewpoints (April 1992).

45. The following is based on correspondence with Gerard Stoddard, president, unless otherwise noted.

46. Fire Island Recollections: Histories of the Island, the Fire Island Association, and its Member Communities (1983).

47. Fire Island Association (1993). "Statement of Policy: Preserving the Barrier Island system; Reconstruction of Storm-Damaged Properties," 1.

48. Ibid.

49. Ibid., 5.

50. Ibid.

51. Gerard Stoddard (1995). "Coastal Forum II: Coastal Policy Implications of Right to Rebuild Questions." *Shore and Beach* 63(1): 27.

52. Ibid., 27.

53. Gerard Stoddard (1992). "Public Policy Issues in Coastal Management Initiatives." Paper presented at the Interdisciplinary Hurricane Conference, Hofstra University (November 6) (mimeo).

54. Stoddard, note 51.

55. Stoddard, note 53.

56. Ibid., 1–2.

57. The following is based on written correspondence and an interview with Jane Ekstein, executive director, Save Our Seashores (March 1995).

58. Statement by Jane Ekstein at the New York State Assembly Hearing on Proposed 1995/1996 State Budget, Plainview, NY (March 16, 1995).

59. National Wildlife Federation (1995). "Takings Bills & Nuisance Exception." Mimeo.

60. Tim D. Searchinger (1993). "Wetland Issues 1993: Challenges and a New Approach," *Maryland Journal of Contemporary Legal Issues* 4(1): 13–72.

61. Roger Pilon (1995). Testimony Before the Senate Committee on Environment and Public Works (June 27).

62. National Wildlife Federation (1995). "The Truth Behind Wetland Horror Stories." Mimeo.

63. Frank I. Michelman (1995). Testimony Before the Senate Committee on Environment and Public Works (June 27).

CHAPTER 5

The Takings Issue and the Regulation of Hazardous Areas

Rutherford H. Platt and Alexandra D. Dawson

> Every man holds his property subject to the general right of the community to regulate its use to whatever degree the public welfare may require it.
>
> Theodore Roosevelt, Speech at Osawatomie,
> August 31, 1910

> The general rule at least is, that while property may be regulated to a certain extent, if regulation goes too far, it will be recognized as a taking.
>
> Oliver Wendell Holmes in
> *Pennsylvania Coal Co. v. Mahon,* 1922

INTRODUCTION

Noncompensatory land use and building regulations applied through state and local laws help to reduce vulnerability to natural disasters in geographic areas of high hazard (e.g., floodplains, coastal areas, steep slopes, and soils subject to seismic failure). Floodplain management regulations (e.g., floodplain zoning and minimum building elevations) have been widely adopted pursuant to the National Flood Insurance Program.[1] Regulations addressing seismic, landslide, erosion, and other geographically specific hazards are employed in certain states and local municipalities.

Today, the entire field of land use regulation, including hazard mitigation regulations, is under broad challenge from property rights interest groups. This chapter reviews the status of constitutional law relating to takings in general, and to hazard mitigation regulations in particular. Two recent Supreme Court decisions involving hazard mitigation, *Lucas v. South Carolina Coastal Council*[2]

131

and *Dolan v. City of Tigard*,[3] will be carefully considered. We argue that despite ominous rhetoric concerning the takings issue, these and other recent cases do not significantly change the long-standing doctrine that land use regulations are necessary and appropriate to mitigate potential public harms, such as those threatened by natural hazards.

THE CONSTITUTIONAL CONTEXT

The control of public hazards through governmental regulation goes back to the earliest recorded history of governmental action, although the early "police power" was more likely to be exercised by armed men than through administrative procedure acts. Governmental constraints were also used to protect limited resources, although (contrary to modern reasoning) such constraints were likely to be called superior property rights. Two examples are the medieval limits on hunting and taking of firewood in the King's forests and the proscriptive designation of larger trees in the colonial U.S. forests as reserved for the King's navy. Because of the Anglo-Saxon passion for land ownership, the foreshore, which under Roman law was incapable of ownership, was in England the property of the Crown.[4]

In 1787, the framers of the U.S. Constitution could scarcely have foreseen that the United States would eventually be a "nation of cities," in the phrase of historian Sam Bass Warner.[5] But reflecting the writings of John Locke, William Blackstone, and other English political theorists, they were acutely concerned with the protection of private property ownership against unreasonable governmental action.[6] It was recognized that government should provide for the "general welfare," as expressed in both the Preamble and in Article I, section 8. This implies the existence of what has become known as the "police power"[7] that is the basis for land use zoning and other noncompensatory public limitations on private freedom.[8]

Private property rights as such were not directly mentioned in the original Constitution but were addressed by the Fifth Amendment in the Bill of Rights, added in 1791.

> No person shall be . . . deprived of life, liberty, or property, without due process of law; nor shall private property be taken for public use without just compensation.

The evil intended to be remedied by this clause apparently related to excesses committed by the British government in the seizure of mostly personal property such as horses and hay. Nevertheless, for almost 200 years the clause was applied exclusively to the taking of real property through eminent domain for public works.[9]

Although the Fifth Amendment applied only to acts of the federal government, the Fourteenth Amendment adopted in 1868 applied portions of the Fifth Amendment to actions of states.

> No State shall . . . deprive any person of life, liberty, or property, without due process of law; nor deny to any person within its jurisdiction the equal protection of the laws.

Although the takings clause is not literally repeated in the Fourteenth Amendment, the latter has long been held to "incorporate" the Bill of Rights including the entire Fifth Amendment to the limitation of state actions. The Fifth and Fourteenth Amendments, as interpreted by courts, thus jointly define the scope of the public power to regulate land use at all levels of governmental authority.

EARLY APPLICATIONS OF THE TAKINGS CLAUSE

At the end of the nineteenth century, the takings clause was limited to actual governmental seizure or occupation of private land where eminent domain had not been exercised to compensate the property owner. In the leading case of *Pumpelly v. Green Bay Co.,*[10] the Supreme Court not surprisingly held that a compensable taking resulted from the permanent flooding of the plaintiff's land by a state-sponsored dam project. Physical occupancy thus became an established basis for the taking doctrine, as extended (perhaps to an extreme) by *Loretto v. Teleprompter Manhattan CATV Corp.,*[11] where a beneficial cable company box not two feet square installed on private property nevertheless was held to require compensation.

On the other hand, in *Commonwealth v. Alger,*[12] the establishment of a harbor line defining the permissible seaward limit of fill or building in Boston Harbor was held valid against a claim that it violated the riparian owner's title to the mean low-water line. The Massachusetts Supreme Judicial Court stated the "settled principle . . . that every holder of property, however absolute and unqualified may be his title, holds it under the implied liability that his use of it may be so regulated, that it shall not be injurious to the equal enjoyment of others having an equal right to the enjoyment of their property, nor injurious to the rights of the community."[13] Freyfogel considers *Alger* to mark "a shift in thinking" whereby the government's "inherent regulatory power" to limit the use of private property was recognized as "distinct from, and not contained by, any pre-existing limitations on a landowner's title."[14]

Until the 1920s, the Supreme Court was extremely tolerant of state regulations over economic activities, including certain forms of land use. In *Mugler v. Kansas,*[15] the court upheld a state law that put an existing brewery out of busi-

Since legislative bodies are constitutionally entrusted with responsibility to recognize and respond to such changing circumstances, courts have long been loathe to overturn the duly adopted acts of legislatures (federal, state, and local). Although the *Pennsylvania Coal* decision provided property owners with a constitutional argument, seldom have they persuaded higher courts that a governmental regulation "goes too far." The presumption of legislative validity was expressed in *Euclid:* "If the validity of the legislative classification for zoning purposes be fairly debatable, the legislative judgment must be allowed to control."[27] The standard for unconstitutionality is not merely that the issue is "fairly debatable" but the provisions in question have "no substantial relation to the public health, safety, morals, or general welfare."[28]

This presumption of legislative validity has meant that the burden of proof lies with property owners to demonstrate that a regulatory law is arbitrary as applied to their properties. As long as the government provides some planning justification, the regulation will normally be upheld. Courts frequently declare that they do not want to be a "super legislature nor a zoning board of appeals."[29] Normally, evidence of a planning basis for a regulatory action will serve to trigger the presumption of validity. In the case of geographically specific natural hazards, a planning basis is provided through maps based on standard scientific methods that indicate the location of areas subject to risk as well as expert testimony, as indicated in several cases discussed in the following section. (The possible effect of the 1994 *Dolan v. City of Tigard* decision on this presumption will be discussed later.)

LOWER COURTS AND HAZARDS

Despite the strong approval given to land use zoning in the 1926 *Euclid* decision, public efforts to apply this tool to floodplains and other hazardous areas were rare before 1960, as were judicial decisions reviewing such regulations. In their classic treatise on floods, hydrologists Hoyt and Langbein offered this gloomy appraisal: "Flood zoning, like almost all that is virtuous, has great verbal support, but almost nothing has been done about it."[30]

In 1954, the California Supreme Court upheld a local "beach recreation district" against a taking claim.[31] The court was influenced by evidence of coastal storm hazards including erosion and flooding pertaining to plaintiff's shoreline property. The precedent value of the case for floodplain regulations, however, was impaired by the apparent purpose of the ordinance, which was to lower the cost of acquiring the site for public recreation, normally an impermissible use of the police power.[32]

In 1959, a seminal article on floodplain zoning by University of Chicago law professor Allison Dunham[33] cited only a handful of precedents. To guide future courts, Dunham offered three rationales to support regulation of floodprone areas: (1) to protect the unwary from investing in or occupying floodprone property; (2) to protect other riparian landowners (upstream, downstream, or cross-stream) from higher flood levels due to ill-considered encroachment on floodplains by their neighbors; and (3) to protect the community from the costs of rescue and disaster assistance. The first two related to classical nuisance law while the third more closely related to public resource protection.

Dunham's approach bore fruit in a landmark 1972 Massachusetts decision in *Turnpike Realty Co. v. Town of Dedham* (Figure 5-1).[34] In reviewing the application of a town floodplain bylaw to a site bordering the Charles River, the court was impressed by testimony that the site had been submerged under as much as four feet of water in recent floods. Brushing aside the owner's contention that such flooding resulted from improper management of a flood control structure, the court adopted Dunham's three arguments almost verbatim to uphold the basic validity of the bylaw:

> The general necessity of flood plain zoning to reduce the damage
> to life and property caused by flooding is unquestionable. . . .
> The restrictions in the by-law serve to protect not only those
> who might choose to develop or occupy the land in spite of the
> dangers to themselves and their property . . . but also other
> people in the community from the harmful effects of flooding.
> Similarly there is a substantial public interest in avoiding the
> public works and disaster relief expenditures connected with
> flooding.[35]

Although the bylaw was thus held to be facially valid, its application to the plaintiff's property posed the issue as to whether it "went too far" (in the Holmsian sense) and destroyed the value of the site without compensation. The owner claimed a loss in value of 88 percent due to the floodplain restriction, but this estimate was disputed by the town. The trial court had found a "substantial diminution in value of the locus" but stated the standard rule that "mere decrease in value of a particular piece of land is not conclusive evidence of an unconstitutional deprivation of property." The state high court upheld that ruling and made particular reference to a statement by the town's expert witness that " . . . the value of the lowlands was not affected by the Flood Plain District as the nature of the terrain was such as to make the construction of housing on the locus . . . not economically feasible."[36]

Six years earlier, the New Jersey Supreme Court in *Spiegle v. Beach Haven*[37] had similarly upheld a challenged beach erosion setback line (Figures 5-2 and 5-3). The court offered the common sense perception that the regulation merely

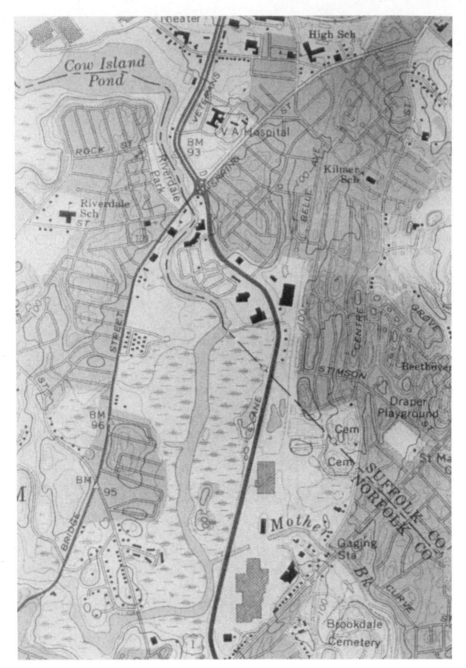

FIGURE 5-1 The scene of the *Turnpike Realty* case, a landmark in floodplain management law. The site in question lay in the wetland just to the left of the heavy dark line (Route 1). (U.S. Geological Survey)

reflected the natural limitations of the site and did not therefore reduce its value, which was already reduced by nature:

> Plaintiffs failed to adduce proof of any economic use to which the property could be put. The borough, on the other hand, adduced unrebutted proof that it would be unsafe to construct houses oceanward of the building line . . . because of the possibility that they would be destroyed during a severe storm—a result which occurred during the storm of March 1962. . . . *Such regulation prescribed only such conduct as good husbandry would dictate that plaintiffs should themselves impose on the use of their own lands.*[38]

A more recent decision by a Federal District Court in *McNulty v. Town of Indialantic*[39] similarly upheld a local/state beach setback regulation in light of obvious natural hazards. McNulty, the owner of undeveloped property "consisting primarily of sandy beach and dune vegetation," was denied a building permit under state and local regulations that established a setback line 50 feet landward of the mean high-water line. All of the plaintiff's property lay seaward of that line. In upholding the permit denial against a taking claim, the court extensively relied on expert testimony predicting "further erosion of the dune by wind currents underneath the structure and as a result of the destruction of dune vegetation, which would be deprived of sunlight and rainfall."[40] In words calculated to gladden the hearts of coastal managers, the court held:

> The town appropriately prohibits construction inconsistent with [state coastal management] goals by exercise of its police power. Its restriction of building in the coastal conservation area and its denial of a variance for construction of a 12-unit condominium complex over the dune and sandy beach is substantially related to the advancement of the legitimate interest of public safety and welfare.[41]

Concerning inland floodplain regulation along streams, a 1991 Illinois case strongly endorsed prohibition of new construction within the 100-year floodway.[42] Severe flooding struck the Chicago metropolitan area in 1986 and again in 1987, transforming local urban streams into dangerous torrents that damaged thousands of homes and businesses. Presidential disaster declarations were declared in both years and the state established a governor's flood control task force to study the causes and recommend legal changes. The Illinois Legislature acted on the commission's report to enact a floodplain management law that prohibited all new residential construction in the 100-year floodway in the Chicago metropolitan area.

harm" in Hurricane Hugo, which ravaged the coast, including the Lucas property, in 1989. The court might have taken judicial notice of that disaster event, just as the New Jersey Supreme Court referred to the 1962 Ash Wednesday Storm in upholding the beach setback law in *Spiegel v. Beach Haven.*[67]

The Supreme Court Decision

When accepted for review by the U.S. Supreme Court, the *Lucas* case quickly became a national test case on property rights versus environmental protection, as reflected in a *Boston Globe* editorial:

> The case has far-reaching implications for the enforcement of regulations concerning everything from billboards to wetlands, as well as the coastline. Environmentalists fear that if the court decides in Lucas' favor, virtually every environmental restriction placed on the use of property will be considered a taking, thus making environmental protection too expensive.[68]

The case was thus viewed as a referendum on the environmental movement, writ large, and the specific concern of coastal hazard mitigation was obscured in the cross-fire of "friend of the court" briefs on broader issues.[69]

The U.S. Supreme Court majority focused on the economic impact on the private owner of the permit denial, but not of the potential economic and social impacts of coastal erosion and flooding on the public. In a 6 to 3 decision, it reversed the state ruling, holding that where a regulation "denies all economically beneficial or productive use of land,"[70] it is a "categorical" compensable taking, equivalent to a physical invasion of the property by governmental action. Justice Scalia, in his majority opinion argued that the need to compensate for "total takings" could not be avoided by merely reciting harms that the regulation would prevent. Moreover, Scalia selectively quotes from purposes of the act relating to tourism, habitat preservation, and recreation (which normally cannot be provided through regulation) with no mention of its hazard mitigation objectives:[71] "One could say that imposing a servitude on Lucas's land is necessary in order to prevent his use of it from 'harming' South Carolina's ecological resources; or, instead, in order to achieve the 'benefits' of an ecological preserve."[72]

In an awkward distinction, Scalia held that even for "total takings," compensation would not be required where a regulation merely reflected a state's "background principles of nuisance and property law."[73] This distinction is ridiculed by Rubenfeld: "*Lucas* begins by telling judges why traditional nuisance reasoning is unworkable in takings law, and it ends by telling them that they must apply *only* traditional nuisance reasoning when they decide total-loss taking cases."[74]

Ironically, in reaching for some examples of such "background principles of nuisance and property law," Scalia alludes to *flooding*—specifically, the flooding of a neighbor's land through the filling of a lake bed.[75] But although filling along a tidal coastline will not similarly heighten flooding on neighboring land, a house may collapse and thereby damage neighboring houses with loose debris.[76] And someone must clean up the debris. Scalia seems uninformed or uninterested in the potential effects of coastal erosion and flooding on structures such as those Lucas proposed to build. This was not for lack of scientific information: The state's brief to the Supreme Court discussed coastal hazards and coastal zone management in detail.[77] But nowhere does the majority discuss the National Flood Insurance Program, the Coastal Barrier Resources Act, or the Coastal Zone Management Act, in relation to the South Carolina Beachfront Management Act. As noted by one of the state's legal counsel, Richard J. Lazarus:

> The Court displayed no appreciation of the factors that led South Carolina to conclude that the physical characteristics of Lucas' land made it the wrong place for the construction of a house. The Court ignored the applicability of its own parenthetical acknowledgment that "*changed circumstances or new knowledge* may make what was previously permissible no longer so."[78]

The "changed circumstances" doctrine, which harks back at least to *Euclid v. Ambler Realty,* was given more weight by Justice Kennedy in his concurring opinion: "The common law of nuisance is too narrow a confine for the exercise of the regulatory power in a complex and interdependent society. . . . The State should not be prevented from enacting new regulatory initiatives in response to changing conditions. . . ."[79] Assuming that the omission of "new knowledge" was inadvertent, Justice Kennedy's statement should logically support the validity of laws like the South Carolina Beachfront Management Act that are based on a thorough study of the real-world problem by a competent study commission.[80]

In furious dissent, Justice Blackmun wrote: "Today the Court launches a missile to kill a mouse."[81] Citing *Keystone* among other cases, Blackmun argued:

> These cases rest on the principle that the State has full power to prohibit an owner's use of property if it is harmful to the public. "[S]ince no individual has a right to use his property so as to create a nuisance or otherwise harm others, the State has not 'taken' anything of value when it asserts its power to enjoin the nuisance-like activity" [*Keystone*] . . . It would make no sense under this theory to suggest that an owner has a constitutionally protected

right to harm others, if only he makes the proper showing of economic loss.[82]

Ultimately, Lucas was paid for the lots and the state offered them for sale with permits to recoup its costs. A large house has been built on one of them. Subsequently, the site was experiencing renewed threat from erosion as the inlet on which it fronts continues to shift (Figure 5-4).

Reaction to *Lucas*

Amid the deluge of legal commentaries on the *Lucas* decision,[83] there is little discussion of the hazard element of the South Carolina Beachfront Management Act, evidently because the U.S. Supreme Court majority never addressed it. Since Lucas only attacked the act on the ground of its economic effect on him ("total taking"), not on its merits, the majority was unwilling to look at the statute itself, except with disparaging comments on tourism and ecology. According to Lazarus, Lucas wanted a "per se" rule: "loss of all value equals a taking requiring compensation."[84] But the Court was not willing to go that far. No

FIGURE 5-4 The scene of the *Lucas v. South Carolina Coastal Council* case. The state's defeat in the U.S. Supreme Court eventually led to the construction of the house on the right on one of the Lucas lots. The beach scarp indicates the inlet is shifting toward the houses again. (Photo by R.H. Platt)

one seems even to have argued the engineering issues so commonly discussed in permit litigation: Why not let Lucas build up on pilings, as permitted under federal regulations? The National Flood Insurance Program was never mentioned in the decision, although it is common in these cases to look at whether a regulation "goes too far" in the sense of imposing more of a burden than is required to get the public benefit desired.[85]

Perhaps the Court did not perceive any hazard or felt that since other homes had been permitted on adjoining property, the hazard was somehow eliminated. It cited *Annicelli v. Town of South Kingston*[86] where a taking was recognized in the denial of permission to build a home on beachfront property in a developed area, as though that somehow settled the question of whether the activity was truly hazardous.

The struggles of courts to distinguish between immediate hazard to individuals or the public (nuisance) and overall need to preserve fragile ecosystems has been discussed by Joseph L. Sax.[87] He concludes that Justice Scalia believed that South Carolina was trying to keep the beach area in its natural condition for tourism and natural habitat without paying for it and decided to send a hard message to legislators about restrictions based on that motivation. According to Sax, the *Lucas* opinion is based on the following reasoning: that requiring land to be left in its natural condition is in "fundamental tension with the traditional goals of private property law" and that if disruption of natural conditions is viewed as harmful, "states could then exercise their police power to maintain such conditions and thereby eliminate property's economic value."[88]

In this sense, *Lucas* is a direct attack on *Just v. Marinette County*[89] and the cases that followed it. *Just* stated a widely cited (and widely disputed) principle that a landowner "has no absolute and unlimited right to change the essential character of his land so as to use it for a purpose for which it was unsuited in its natural state and which injures the rights of others."[90] But one need not subscribe to the "environmental utopia" of *Just* in order to recognize, in the words of the New Jersey court in *Speigel,* that preventing new construction on an eroding and storm-threatened coast is simply "good husbandry."

Calculating the Extent of "Taking"

As Rubenfeld notes, the *Lucas* decision is full of ambiguities.[91] On its face, however, *Lucas* simply announces that a governmental regulation that deprives the landowner of substantially all economic benefit of his or her land constitutes a "taking" of that property under the 5th and 14th Amendments to the U.S. Constitution, unless the regulation is founded on traditional state doctrines of property or nuisance such that a neighbor or other person directly affected could bring a legal action to stop the activity. Unfortunately, the decision did not go on to say how compensation should be figured, or whether the government had to take title to the land if the landowner wanted to keep it and get some money,

interest vested in the owner, as a matter of state property law, and
not within the power of the state to regulate under common law
nuisance doctrine.[113]

The court hailed the third element as a "sea change" in judicial doctrine
regarding the taking issue that replaces:

> . . . ad hoc balancing decisions, balancing private property
> rights against state regulatory policy [with] one in which state
> property law, incorporating common law nuisance doctrine, con-
> trols. This sea change removed from regulatory takings the
> vagaries of the balancing process, so dependent on judicial per-
> ceptions with little effective guidance in law. It substitutes
> instead a referent familiar to property lawyers everywhere, and
> one which will have substantial (though varying from state to
> state) likelihood of predictability for both property owners and
> regulators.[114]

This laudatory view of Scalia's "common law nuisance and property law"
exception to the "total taking" doctrine is at odds with Justice Blackmun's dissent in *Lucas* quoted earlier as well as the views of many commentators. For
example, Eric Freyfogel writes:

> In its search to avoid such fuzziness, the Court in *Lucas* set aside
> the harm-benefit test. The better approach, the Court announced,
> was to look to the past, to ownership's "background principles."
> These principles would supply the foundation. Yet it takes no
> more than superficial scratching to reveal the flaws in this line of
> reasoning. The dominant principle of ownership law is sic utere,
> the underlying norm of nuisance law. As applied, nuisance law is
> little more than the harm-benefit test.[115]

The court held that the burden of proof is on the government to show that a
law abates a common law nuisance so as to fall within the exception. In the
absence of such proof, a regulatory taking was found.[116]

The "denominator problem" is the most interesting aspect of the *Loveladies
Harbor* decision. Justice Scalia in the *Lucas* decision distinctly announced that he
was not interested in hewing to the *Penn Central* rule of considering all the
landowner's properties in the vicinity. Presumably he would not be interested in
past sales and development of portions of the same land. Severance of buildable
from less buildable portions of tracts, rather than "partial takings" as such, is

likely to be the next step after *Lucas*—an attack on the "denominator" against which total loss is established.[117]

This approach to takings offers dangers and opportunities. The real danger is that developers will be encouraged to develop the "easy" or safest parts of property first, then plead a taking when all that is left is floodplain, wetlands, unstable hill slope, and so on. The solution is to develop a "single project" approach by which the whole property is agreed in advance to be a single entity of value, so that the "easy" and "not easy" parts are perpetually linked (see below). This problem arose in Massachusetts's first wetlands taking case, *Commissioner of Natural Resources v. S. Volpe & Co.,*[118] where the issue of the loss of value to the wetlands became severed from the value the wetlands probably had as ancillary to developable uplands.[119]

Massachusetts's first post-*Lucas* land use decision, *Lopes v. City of Peabody,*[120] also demonstrates the difficulties of enforcing wetlands laws against a "total takings" claim where the restricted land is viewed separately from the part already developed. Lopes bought a quarter-acre building lot totally subject to a wetlands/floodplain building ban for $16,000. The trial court had upheld the ban as not requiring compensation, presumably because of flooding problems. The state Appeals Court agreed, and the state Supreme Judicial Court (SJC) denied further review. The U.S. Supreme Court[121] remanded the case for reconsideration in the light of *Lucas*.

Upon reconsideration, the SJC pointed out that at trial the parties had stipulated that Lopes "is unable to use his land as a result of the revision of the Zoning Ordinance."[122] This unfortunate stipulation removed any consideration of residual uses from the case. The court thus remanded the case to the Land Court to determine if this denial of "all economically beneficial use" constituted a taking or if the restriction was "independently justified by other principles of land use law. . . ."[123] The Land Court duly found there would have been a taking under the *Lucas* doctrine, except that it was able to redraw the restrictive elevation line so that the lot became buildable. The Land Court also found a probable interim taking under the *First English* doctrine.[124]

A further interesting point emerging from the *Lopes* decision is whether it mattered that Lopes bought his property after it was subject to the wetlands zoning restriction. The court ruled that "a purchaser of land subject to the restriction at the time of his purchase has every right to challenge the continued application of the restriction."[125] However, the court, without explanation, also ruled that this factor might well affect significantly the amount of compensation due Lopes.[126] (This issue did not arise for Lucas, who acquired his property before the S.C. Beachfront Management Act was passed.)

In 1996, the Massachusetts SJC on its own motion reviewed a *Lucas*-based challenge to a local floodplain zoning regulation in *Leonard v. Town of Brim-*

field.[127] The plaintiff sought approval to subdivide a 16-acre tract within a mapped floodplain area. The town allowed construction on six acres and limited the rest to agricultural use. In response to a claim of a regulatory taking as to the nonbuildable portion, the SJC decided that plaintiff bought the entire site with notice of the floodplain restrictions that were in effect and therefore could not claim that her "investment-backed expectations" to develop the entire area were reasonable. Citing Scalia's exception in *Lucas,* it held " Because she purchased the property subject to the restrictions on building in a floodplain, she may not complain about the loss of a right she never acquired."[128] The court viewed the whole tract as an entirety, and held that her right to build on part of the site would provide adequate economic return to avoid a regulatory taking by the town.

Dolan v. City of Tigard

As of January 1999, the most significant taking issue/property rights case since *Lucas* has been *Dolan v. City of Tigard.*[129] The case concerned various land exactions imposed on a property owner by a local government as conditions for a development permit. Specifically, the plaintiff who sought to enlarge her hardware store was required to dedicate to the city a portion of her land within a 100-year floodplain, plus an additional 15-foot strip adjoining the floodplain to be used as part of the city's pedestrian and bikeway system. These exactions amounted to about 15 percent of her total land area. Upon challenge by the owner that these exactions took a portion of her property without compensation, the city was upheld by the Oregon Supreme Court, which was in turn reversed by the U.S. Supreme Court. As with *Lucas, Dolan* attracted a "mountain of briefs" (in the words of Justice John Paul Stevens) and hyperbole in the press.[130]

Except for its outcome, the majority opinion in this 5-to-4 decision is supportive of the efforts of planners and hazard mitigators. Writing for the majority, Chief Justice William Rehnquist states in dictum:

> Undoubtedly the prevention of flooding along Fanno Creek and the reduction of traffic congestion in the central business district qualify as the type of legitimate public purposes we have upheld.
> . . . *It seems equally obvious that a nexus exists between preventing flooding along Fanno Creek and limiting development within the creek's 100-year floodplain.*[131]

The majority had difficulty, however, with the extent of the burden imposed on the private owner, especially in the case of the bikeway exaction that involved public access, usually a forbidden outcome of noncompensatory regulations. The Court explicitly invoked and expanded on *Nollan,*[132] where a beach access easement was rejected. Rehnquist required the defendant City to demonstrate what it called "rough proportionality" between the burden on the property owner and the public to be avoided: "No precise mathematical calculation is required, but the city must make some sort of individualized determination that the required dedication is related both in nature and extent to the impact of the proposed development."[133]

In dissent, Stevens, joined by Blackmun and Justice Ruth Bader Ginsburg, argued that this "rough proportionality" test places the burden of proof on the city, reversing the long-standing presumption of validity extended by courts to the regulatory actions of local governments. "Rough proportionality," he stated, also is difficult to satisfy in relation to such statistically uncertain phenomena as flooding and traffic congestion.

> In our changing world one thing is certain: Uncertainty will characterize predictions about the impact of new urban developments on the risks of floods, earthquakes, traffic·congestion, or environmental harms. When there is doubt concerning the magnitude of those impacts, the public interest in averting them must outweigh the private interest of the commercial entrepreneur.[134]

With a shift of one justice, this could have been the majority view and thus the "law of the land."[135]

Despite the apocalyptic reaction of the press, *Dolan*, like *Lucas*, need not be cause for alarm to hazard mitigators if it is read carefully. The Court's primary concern was certainly the issue of public access across private property (albeit commercial in nature). Without the bikeway issue, the floodplain restriction probably would never have reached the High Court. It is arguable that the "rough proportionality" test issued by the majority was meant to apply to the bikeway exaction, not the floodplain. Protection of the latter, of course, could have been accomplished by straight floodplain regulation without any dedication to public use. It might benefit the property owner to let the city maintain the floodplain, as in clearing brush and removing the inevitable shopping carts and other junk.

CONCLUSION

It has long been the advice of environmental publications that regulation should focus on hazard prevention where applicable and avoid emphasis on habitat protection, public recreation, or esthetic values.[136] It is therefore critical that when hazard mitigation is the major purpose of a regulation, it should be so stated and offhand references to "ecology" and similar terms should be avoided. The *Loveladies Harbor* decision, for instance, disparaged the South Carolina Beachfront Management Act as "state-enacted legislation aimed at preserving and protecting the fragile ecology of the beach and dune area."[137] In *Burrows v. City of Keene,*[138] the New Hampshire Supreme Court, which has shown great sympathy to wetlands protection, blasted a zoning effort by the city that seemed directed entirely at preserving open space that had no hazards on it.

If Professor Sax is right, then the *Lucas* decision, in spite of a stated rejection of the doctrine, is really based in part on the old "harm versus benefit" distinction: Harmful hazards can be regulated without compensation through the police power, while public benefits, such as parks or open space, must be acquired through governmental purchase or condemnation. That these distinctions make less and less sense in a highly populated and ecologically interrelated world, seems to be irrelevant to some courts.

It is a curious anomaly that technological requirements, no matter how costly to the developer, are not viewed as "takings" while limitations on land use may be so viewed. Minimum elevations of coastal buildings required under the National Flood Insurance Program are seldom challenged, and reported decisions involving them are not known to the authors. Nor have we discovered a case in which a court required compensation for a development ban based on the proven unsuitability of an area for sewage disposal. Improper disposal of human waste is the classic nuisance.[139] Perhaps in time, floods, hurricanes, and earthquakes will be viewed as categorical threats to society as well, with reasonable mitigative regulations given an overwhelming presumption of validity.

It would be wise to avoid the "last remnant" situation that decided the *Loveladies Harbors* case, through a regulatory framework that requires consideration of the whole parcel undergoing development as a single entity. This is already in place in the Massachusetts state review of wetlands alteration under Section 404 of the Clean Water Act, the federal wetlands law. Under this system, for example, a residential or industrial subdivision is reviewed as a single project, in terms of wetlands-alteration limits, even though the lots may be developed over several years. If inserted into the chain of title, via permit recording, such a single-project declaration could act as the sort of property rights restriction that Scalia recognized as an exception to the takings doctrine.

Finally, a variance or limited exception provision is desirable to avoid a total taking of property value where small areas are involved. The Massachusetts Wetlands Protection Act regulations[140] have provided both since 1983. The most common exception is a provision that 5000 square feet of wetlands may be altered on any property by permission of the local conservation commission or, on appeal, by the state Department of Environmental Protection. The variance provision, basically designed for public works, rests with the Commissioner of DEP. An unlimited variance is, however, subject to abuse and should be omitted if possible. This approach is supported by Strong, Mandelker, and Kelly who recommend that land use decisions be framed as "quasi-judicial" proceedings for individual owners or parcels to promote flexibility and provide a better record of the public planning justification in case of challenge.[141]

These "safety valves" may also be helpful in avoiding the ominous effects of the *First English* decision requiring compensation for "temporary takings," apparently measured as loss of value during the pendency of a restriction later found invalid. Before this decision, there was little risk for state and local bodies to employ strong regulation, for if such a regulation was found confiscatory as applied to any landowner, the regulatory agency could simply remove the offending restriction without paying for any damages suffered. The *First English* case has surely had a chilling effect on legislative bodies considering regulations. However, it is possible that, should a limited exception or variance be rejected by an applicant seeking instead an absolute taking finding, a high court could be persuaded at least to limit damages under the doctrine that "no compensable taking is caused by fluctuations in value during the process of governmental decision making, absent extraordinary delay."[142] The Massachusetts Supreme Judicial Court followed this doctrine in *Wilson v. Commonwealth*[143] in finding no taking where protracted negotiations between state agencies and coastal landowners regarding installation of seawalls were interrupted by a storm that in effect "took" the subject houses. The court cited *Lucas* as not in point because this case could involve adverse effects on property of others—a nuisance argument, but one that may be dictum since the decision was based on the impact of regulatory delay.

Until further judicial shoes fall, it is hard to tell whether *Lucas* represents a trend or an anomaly. In any case, it is clear that many courts will continue to favor regulations that protect human lives and health and property rather than ecological integrity. Regulations therefore can and should be written and enforced with the mental image of Scalia's watchful eye. It might also be wise to stay in sympathetic state courts, when a choice is offered.

Ultimately, the two adages quoted at the beginning of this chapter should be kept in mind. Property owners do not have a right to use their property to the detriment of themselves, their neighbors, their successors, or the general public. But the public does not have the right to compel property owners to bear a disproportionate burden toward the fulfillment of public goals. In the case of natural hazards mitigation regulations, the balancing act posed by the two propositions must be resolved through incorporation of the best available scientific knowledge into legislative, regulatory, and judicial processes.

Notes

1. Raymond J. Burby, Steven P. French, and others (1985). *Flood Plain Land Use Management: A National Assessment.* Boulder: Westview Press; James M. Holway and Raymond J. Burby, "Reducing Flood Losses: Local Planning and Land Use Controls" *Journal of the American Planning Association* 59(2): 205–216.

2. 112 (1992) S.Ct. 2886.

3. 114 (1994) S.Ct. 2309.

4. Eric T. Freyfogle (1995). "The Owning and Taking of Sensitive Lands," 43 *UCLA Law Review* 77–138.

5. Sam Bass Warner (1966). *Planning for a Nation of Cities.* Cambridge, Mass.: MIT Press.

6. James M. McElfish, Jr. (1994). "Property Rights, Property Roots: Rediscovering the Basis for Legal Protection of the Environment" 24 *ELR News and Analysis* 10231–10249.

7. Ernst Freund (1904). *The Police Power: Public Policy and Constitutional Law.* Chicago: Callaghan.

8. Rutherford H. Platt (1996). *Land Use and Society: Geography, Law, and Public Policy.* Washington, D.C.: Island Press, Chapters 7 and 8.

9. Jeb Rubenfeld, "Usings" 102 *Yale Law Journal* 1077 (1993).

10. 80 U.S. (13 Wall.) 166 (1871).

11. 458 U.S. 419 (1982).

12. 61 Mass. 53 (1851).

13. Ibid., at 84–85.

14. Freyfogle, note 4, 103.

15. 123 U.S. 623 (1887).

16. 214 U.S. 91 (1909).

17. 239 U.S. 394 (1915).

18. 260 U.S. 393 (1922).

19. Ibid., at 413.

20. 260 U.S., at 415.

21. Ibid. (emphasis added).

22. Ibid., at 417 (emphasis added). For a recent reappraisal of the *Pennsylvania Coal* case, including an account of what happened to Scranton after the coal company won (nothing), see William Fischel (1995). *Regulatory Takings: Law, Economics, and Politics.* Cambridge: Harvard University Press, Ch. 1.

23. 107 S.Ct. 1232 (1987).

24. Ibid., at 1243.

25. 272 U.S. 365 (1926).

26. Ibid., at 386–387.

27. Ibid., at 388.

28. Ibid., at 395.

29. *Construction Industry Assn. of Sonoma County v. City of Petaluma* 522 F.2d (1975), 897.

30. William G. Hoyt and Walter B. Langbein (1955). *Floods.* Princeton, N.J.: Princeton University Press, at 95.

31. *McCarthy v. City of Manhattan Beach* 264 P.2d 932 (1954).

32. Also see *Burrows v. City of Keene* 432 A.2d 15 (1981).

33. Allison Dunham (1959). "Flood Control Via the Police Power," 107 *University of Pennsylvania Law Review* 1098–1132.

34. 284 N. E. 2d 891, *cert. den.* 93 S.Ct. 908 (1973).

35. Ibid., at 899.

36. Ibid., at 900.

37. 218 A.2d 129 (1966).

38. Ibid., at 137 (emphasis added).

39. 727 F.Supp. 604 (M.D.Fla. 1989).

40. Ibid., at 607.

41. Ibid.

42. *Beverly Bank v. Illinois Department of Transportation* 579 N.E.2d 815 (1991). The "100-year floodway" referred to in this case is "the entire area which would be at or under the water level . . . in the worst storm likely to occur in 100 years" (ibid., at 816). The term is thus synonymous with the more commonly used "100-year floodplain," which is any area bordering a stream, lake, or marine coast with a risk of being flooded at least once a century, or a chance of 1 percent of being flooded in any given year. Either term is misleading since the actual risk within the floodplain is much higher for locations closer to the river channel or coastline: Only the outer fringe has a risk as low as 1 percent per year. Also in small urbanizing watersheds, land use changes upstream may drastically increase flood risk in downstream areas and thus expand the area of the "100-year flood." See Interagency Floodplain Management Review Committee (1994), *Sharing the Challenge: Floodplain Management into the 21st Century*. Washington, D.C.: U.S. Government Printing Office, 59–60.

43. Ibid., at 822.

44. 193 A.2d 232 (1963).

45. *A.M.G. Associates v. Springfield Township* 319 A.2d 705, at 711. Other cases finding hazard or wetland regulations to inflict takings include *Dooley v. Town Plan and Zoning Commission of Fairfield* 197 A.2d 770 (Conn., 1964); *MacGibbon v. Board of Appeals of Duxbury* 200 N.E.2d 254 (Mass., 1964); *State v. Johnson* 265 A.2d 711 (Maine, 1970); and *Annicelli v. Town of South Kingstown* 463 A.2d 133 (R.I., 1983).

46. 82 S.Ct. 987 (1962).

47. Ibid., at 989.

48. 98 S.Ct. 2646 (1978).

49. See *City of Hollywood v. Hollywood, Inc.* 432 So.2d 1332 (Fla., 1983) where a lower court upheld a municipal TDR scheme that downzoned an oceanfront strip and offered higher density if the owner built a short distance landward from the shore. The court characterized the beachfront site as "the last unspoiled beach area on the Gold Coast, a veritable Shangri-la in an otherwise endless Himalayan mountain range of cement to the south." (432 So.2d, at 1335). Also see Ann Louise Strong, Daniel R. Mandelker, and Eric Damian Kelly, "Property Rights and Takings," *Journal of the American Planning Association* 62(1): 5–16, at 14.

50. 432 So.2d, at 2662–2663. See also *Keystone Bituminous Coal Assn. v. DeBenedictis* 107 S.Ct 1232 (1987), which held that the portion of coal required to be left in the ground could not be viewed separately from the total corpus of rights that a coal company owned in a particular tract for purposes of takings jurisprudence.

51. 107 S.Ct. 2378 (1987).

52. *First English Lutheran Evangelical Church v. County of Los Angeles* 258 Cal. Rptr. (1989), 930.

53. 258 *Cal. Rptr.*, at 904. But see Richard Epstein (1987), "Takings: Descent and Resurrection," *Supreme Court Review* 1. Epstein argued that no matter how trivial the compensable loss, interim damages should have been paid in *First English*, which he predicted would become "a permanent fixture in every land use battle in the United States" (p. 31).

54. 107 S.Ct 1232 (1987).

55. 107 S.Ct. 3141 (1987).

56. Ibid., at 1236–7.

57. Note 39.

58. 107 S.Ct., at 3150.

59. Rutherford H. Platt, "Review of William A. Fischel, *Regulatory Takings: Law, Economics, and Politics.*" In *Issues in Science and Technology* 12(2): 85–89.

60. Rathkopf (1991). *Law of Planning and Zoning*, sec. 7.07, 7–85.

61. 112 S.Ct. 2886 (1992).

62. S.C. Code, secs. 48-39-250 et seq.

63. Rutherford H. Platt (1992). "An Eroding Base," 9 *The Environmental Forum* 10–15.

64. *Lucas v. South Carolina Coastal Council* 404 S.E.2d (1991), 895.

65. Ibid., at 898 (emphasis added).

66. Ibid., at 899.

67. Note 39, supra. Hurricane Hugo led to a weakening of the Beachfront Management Act by the South Carolina legislature to allow special permits for construction seaward of the erosion baseline. Although Lucas probably could have qualified for such a permit, he pursued his original challenge, which raised the issue of mootness in the view of some of the justices. (See concurring opinion by Justice Kennedy.) But the majority nevertheless chose to review the case on its merits since the state court had done so. See Rutherford H. Platt, Timothy Beatley, and H. Crane Miller (1991). "The Folly at Folly Beach and Other Failings of U.S. Coastal Erosion Policy," *Environment* 33: 6–19; 25–32.

68. "Private Rights, Public Benefit," *The Boston Globe* (March 5, 1992), 1.

69. One brief submitted by the National Audubon Society and the National Resources Defense Council and others, including Rutherford Platt and Orrin Pilkey, the noted Duke University coastal geologist, discussed coastal hazards and sea level rise as justification for the South Carolina Beachfront Management Act.

70. 112 S.Ct., at 2893.

71. Ibid., at 2898, fn. 11.

72. Ibid., at 2898.

73. Ibid., at 2901.

74. Rubenfeld, note 9, 1094.

75. 112 S. Ct., at 2900.

76. For example, as of August 1996, the beachfront of the Fire Island National Seashore on the South Shore of Long Island, New York, was strewn with debris from the destruction of homes in adjoining private residential communities during the previous winter. The National Park Service has no funds to collect and dispose of the debris. (Personal communication, Diane Abell, National Park Service planner, August 27, 1996.)

77. Respondent's Brief on the Merits submitted to the U.S. Supreme Court in *Lucas v. South Carolina Coastal Council* (Case No. 91-453).

78. Richard J. Lazarus, "Putting the Correct Spin on *Lucas*" 45 *Stanford Law Review* 1411 (1993), at 1421–1422 (emphasis added).

79. 112 S. Ct., at 2903.

80. Justice Kennedy, however, concurred in the decision to remand to the state court. The *Lucas* case was remanded to the state court for a determination as to whether the "background nuisance or property law" exceptions applied. The state court held they were not and that an interim taking had occurred (*Lucas v. South Carolina Coastal Council* 424 S.E.2d 484 (S. C., 1992).

81. 112 S. Ct., at 2904.

82. 112 S. Ct., at 2912.

83. See, e.g., "Symposium on *Lucas v. South Carolina Coastal Council*" 45 *Stanford Law Review* 1369 (1993); Glenn P. Sugamel, "Takings Issues in Light of *Lucas v. South Carolina Coastal Council:* A Decision Full of Sound and Fury Signifying Nothing" 12 *Virginia Environmental Law Journal* 439 (1993); Frank I. Michelman, "Property, Federalism, and Jurisprudence: A Comment on *Lucas* and Judicial Conservatism," 35 *William and Mary Law Review* 301 (1993); Rutherford H. Platt, "An Eroding Base" 9 *Environmental Forum* 10 (1992); and Joseph L. Sax, "Property Rights and the Economy of Nature: Understanding *Lucas v. South Carolina Coastal Council,*" 45 *Stanford Law Review* 1411 (1993).

84. Lazarus, note 78, 1415.

85. See, e.g., *Nollan*, note 55.

86. 463 A.2d 133 (R.I., 1983).

87. Joseph L. Sax (1993). "Property Rights and the Economy of Nature: Understanding *Lucas v. South Carolina Coastal Council,*" 45 *Stanford Law Review* 1433.

88. Ibid., at 1441–1442.

89. 201 N.W.2d 761(Wis., 1972).

90. Ibid., at 768.

91. Rubenfeld, note 9.

92. Richard A. Epstein (1993). "*Lucas v. South Carolina Coastal Council:* A Tangled Web of Expectation," 45 *Stanford Law Review* 1369.

93. 114 S. Ct. 2309 (1994).

94. See note 92 supra.

95. 63 *Law Week* 4665 (1995).

96. 112 S. Ct., at 2914, fn. 19.

97. William M. Fisher III (1993). "The Trouble with *Lucas*" 45 *Stanford Law Review* 1393.

98. Citing *Bowditch v. Boston* 101 U.S. 16 (1880).

99. 276 U.S. 272 (1928).

100. Sax, note 87.

101. Note 34.

102. Strong, Mandelker, and Kelly, note 49.

103. 854 P. 2d 449 (Or., 1993).

104. Ibid., at 942.

105. 114 S.Ct. 1332 (1994).

106. Ibid., at 1335, fn. 4.

107. 18 F.3d 1560 (Fed. Cir. 1994).

108. 28 F.3d 1171 (Fed. Cir. 1994).

109. 528 A.2d 453 (Maine, 1987).

110. 18 F.3d, at 1565.

111. 28 F.3d, at 1175.

112. Ibid.

113. Ibid., at 1179.

114. Ibid.

115. Freyfogel, note 4.

116. 28 F.3rd, at 1183.

117. Fisher, note 97, at 1402.

118. 206 N.E.2d 666 (1965).

119. But see S. *Kemble Fischer Realty Trust v. Board of Appeals of Concord* 402 N.E.2d 100 (1980) where the state appeals court found residual value in unbuildable floodplain land because it served to make up the required lot size for upland lots sloping into the floodplain.

120. 629 N.E.2d 1312 (1994).

121. *Lopes v. City of Peabody* 112 S. Ct. 2886 (1992).

122. 629 N.E.2d, at 1314.

123. Ibid., at 1316.

124. Essex Misc. #139663, March 1995.

125. 629 N.E.2d, at 1315.

126. Ibid., fns. 7 and 8.

127. 666 N.E.2d 1300 (1996).

128. Ibid., at 1303.

129. 114 S.Ct. 2309 (1994).

130. *The New York Times* (June 25, 1994) reported the decision with a full page of coverage under the headline: "High Court Limits the Public Power on Private Land . . . Opinion by Rehnquist Curbs Environmental and Other Land Use Measures."

131. 114 S.Ct., at 2318 (emphasis added).

132. Note 55.

133. 114 S.Ct., at 2319–2320.

134. Ibid., at 2329.

135. Rutherford H. Platt (1994). "Parsing *Dolan,*" 36 *Environment* 4.

136. See, e.g., Dawson and Zielinski (1991). *Environmental Handbook For Massachusetts Conservation Commissioners* (7th ed.), ch. 10.

137. 28 F.3rd, at 1178.

138. 432 A.2d 15 (1981).

139. It will be interesting to see if the ban newly imposed by Massachusetts Sanitary Code ("Title 5") on leachfields in regulatory floodways and coastal high-hazard areas (as defined by FEMA) can be successfully attacked as nonscience-based and therefore effecting takings.

140. 310 CMR 10.24, 53, and 58.

141. Note 51, supra.

142. 107 S.Ct., at 2388.

143. 597 N.E.2d 43 (Mass., 1992).

PART III

CASE STUDIES

At this point, the discussion shifts from a national policy perspective to a regional and local focus. Issues of federal disaster policy are now approached from a grass-roots, empirical perspective. The next three chapters present in-depth case studies of three different hazards in three regions of the United States. Chapter 6 addresses the history and present controversy over beach erosion at Fire Island, New York. Chapter 7 provides a vignette of the Midwest Flood of 1993, as experienced in St. Charles County, Missouri, at the confluence of the Mississippi and Missouri Rivers. Chapter 8 summarizes natural disaster experience in the Bay Area of California, with particular attention to the Oakland Firestorm of 1991.

The case studies thus survey a broad range of experience in preparing for and recovering from diverse natural hazards in very different geographical and political contexts. The relationship of federal, state, local, and private interests also varies from one setting to another. The NFIP, for instance, is relevant to the Fire Island and Missouri cases, but not to the Bay Area earthquake and fire disasters.

Aside from which combination of federal programs and policies are involved, we seek to examine how the federal government, deliberately or inadvertently, has influenced the outcome of the disaster recovery process. In other words, how has the objective of hazard mitigation been influenced, for better or worse, by the infusion of federal grants, insurance payments, disaster loans, and other forms of assistance. And in the process, have alternative strategies for recovery been ignored in the rush to rebuild. A further issue is the extent to which land use regulations at all levels of government have proven to be helpful to the goal of reducing future losses.

CHAPTER 6

Fire Island:
The Politics of Coastal Erosion

Rutherford H. Platt, David Scherf,
and K. Beth O'Donnell

But look! Here come more crowds, pacing straight for the water,
and seemingly bound for a dive. Strange! Nothing will content
them but the extremest limit of the land. . . . They must get just
as nigh the water as they possibly can without falling in.

Herman Melville, *Moby Dick* (1851)

INTRODUCTION

From Rockaway Inlet in New York City eastward to the head of Shinnecock Bay
at Southampton, Long Island's, south shore is fringed by a 70-mile-long chain of
coastal barriers (Figure 6-1). Nearly half of this distance is spanned by Fire
Island, which extends 32 miles between Fire Island Inlet to the west and
Moriches Inlet to the east. Fire Island forms a narrow ribbon of sand separating
the crashing surf of the open Atlantic Ocean from the tranquil Great South Bay,
a broad and shallow estuary rich with clam flats. The western half of the island,
closer to "the city," is segmented into a series of discrete summer colonies, each
with its own socioeconomic character and boat service from the mainland. In the
absence of any lengthwise road on the island, cars are parked for ten dollars a
night at ferry terminals on Long Island, and summer people ranging from
investment bankers to bartenders tote their luggage, groceries, booze, pets, and
sports gear onto fast passenger ferries for the half-hour crossing to the island.
Once there, the hedonistic rites of summer begin, according to the customs of
each locality.

Thanks to far-sighted conservationists in the 1960s, over three-quarters of the
island, including most of its eastern half, remains fairly natural under the auspices of the Fire Island National Seashore. Two other public parks at its eastern

167

FIGURE 6-1 General location of Fire Island in relation to New York City.
(Source: Matrix Maps)

and western extremities are the only parts of Fire Island accessible to the general
public by bridge and paved road.

Hurricanes, winter storms, and ongoing erosion have taken a huge toll on the
communities of Fire Island. The oceanfronts of the older communities were once
lined with elegant shingle-style summer cottages standing behind a boardwalk,
grassy dunes, and a broad beach (Figure 6-2). Today, virtually every one of those
turn-of-the-century homes has been washed away or moved landward. The
oceanview boardwalks are gone and the "dunes," if present at all, consist of
ridges of bulldozed sand planted with rows of dune grasses and lined with sand
fences to hold them in place. The beach is so narrow in places that no dry sand
remains between the high tide line and the fences. Today, hundreds of homes,
many of them originally two or three rows back from the beach, perch precariously at the water's edge, unprotected by anything except the hope of a federal
renourishment project (Figure 6-3).

This interesting and complex coastal landform is the scene of unceasing controversy locally, in the state capital at Albany, and in Washington, D.C. The
question at issue is who should bear the costs of maintaining the status quo at
Fire Island against the ceaseless onslaught of the Atlantic Ocean. This chapter
outlines the nature of the Fire Island controversy as a microcosm of the contra-
dictory roles of the federal government regarding the protection of seasonal
property from coastal hazards. On the one hand, the federal government is the

FIGURE 6-2 The oceanfront row at Point O'Woods, sometime before the 1938 Hurricane. Cottages are situated landward of boardwalk, dune line, and a broad beach.

FIGURE 6-3 The beachfront at Ocean Bay Park, Fire Island, 1994. High tide line is up to the sand fence and remnant dune. (Photo by R.H. Platt)

prime source of disaster assistance, the rebuilder of damaged communities, and the principal champion of hazard mitigation. On the other hand, the federal government fuels the momentum of building and rebuilding in hazardous coastal areas as the consequence (intended or not) of a variety of incentives, subsidies, and ad hoc projects.

THE PHYSICAL SETTING

Geologically, Fire Island is composed of sand and smaller unconsolidated sediments eroded and deposited from Long Island during the Holocene (post-glacial) epoch. Long Island itself is composed of sediments deposited as moraines and outwash plains from Pleistocene (Wisconsinan) glaciation. Two terminal moraines trend eastward from New York City along the north shore of Long Island (Harbor Hill moraine) and along the center of Long Island to the southeast (Ronkonkoma moraine).[1] These moraines were deposited during the Wisconsinian glaciation and served as a sediment source for the outwash plain that formed from deposits laid down by glacial meltwater along the south shore of mainland Long Island. Headlands at Montauk, which are a segment of the Ronkonkoma moraine, continue to supply sediment to the barrier islands that fringe the mainland.[2]

Like all coastal barriers, Fire Island is dynamic. Over the past several thousand years it has retreated gradually landward through the process of "rollover," the periodic washover of beach material across the barrier toward the bay. Peat deposits recently discovered on the oceanfront of Fire Island originated in freshwater ponds when the island lay far seaward of its present position.[3] The natural rate of recession of Fire Island before the opening of updrift inlets (Shinnecock and Moriches) has been estimated at about one foot per year.[4] Allen and Psuty[5] estimate the average rate of erosion before inlet stabilization at about 0.5 meter (about 1.5 feet) per year with wide variation. Also, Fire Island over time has extended westward as the littoral drift transports beach material from the bluffs and beaches of eastern Long Island toward Fire Island Inlet. The latter has shifted substantially westward in recorded times. Fire Island Lighthouse was constructed in 1858 only 500 feet from the inlet, which is now nearly five miles west of the light.[6]

The width and position of the island is significantly influenced by inlet formation and closure.[7] Tides flowing in and out of inlets lay down large quantities of sediment, creating ebb and flood tidal deltas, respectively seaward and bayward of the inlet. Thus potential beach material is diverted to form deltas at either end of the inlet, until it eventually fills up and closes. The opening of Moriches Inlet in 1931 and Shinnecock Inlet in the 1938 Hurricane intercepted quantities of potential beach sand, thus increasing shore erosion along Fire Island downdrift (west) of those inlets. That effect was perpetuated with the stabilization of those inlets through construction of jetties. The distance to which those stabilized inlets affect erosion rates along Fire Island is currently in dispute. Cashin Associates states: "Active inlet breaches can sometimes cause severe deficiencies in littoral drift and induce increased shoreline erosion for distances that may extend for *several miles* in the down-drift direction."[8] (emphasis added). Property owners in communities as far as 20 miles west of the Moriches Inlet argue that their erosion problems are largely attributable to the jetties at that inlet and a groinfield farther east at Westhampton Beach. Researchers disagree as to the extent of downdrift impact of those structures.

Fresh groundwater underlies the island in a "lens" that is maintained by its ability to float on denser, salty ground water (this principle is known as the Ghyben-Herzberg ratio).[9] The elliptical lens is maintained solely by precipitation. Nonsalt marsh plant species and human habitations are dependent on this freshwater lens.[10]

Ecologically, Fire Island can best be described as fitting into the Atlantic coastal biome—the northeastern oak-pitch pine region.[11] It retains the last substantial tracts of undeveloped coastal barrier habitat to be found on the ocean shorelines of Long Island. The Sunken Forest, a climax maritime forest and unique ecosystem, is protected within the Fire Island National Seashore. The forest has grown up behind and below the crest of the island's dune system. It is a unique forest of trees (American holly, sassafras, tupelo, and shadblow) with an intertwined, dense growth of vines (catbrier, poison ivy, and wild grape).

FIRE ISLAND'S SUMMER COMMUNITIES

Origins

Located some 50 miles from New York City, the largest and wealthiest urban agglomeration in the United States, Fire Island became a prototype coastal resort by the late nineteenth century. It was a destination for persons wanting to experience hunting, sailing, bathing, and civilized relaxation. To this end, Felix Dominy established the Dominy House around 1840 while serving as keeper of the Fire Island Light.[12] Dominy's establishment was the predecessor to the Surf Hotel built at Kismet in 1856, a grandiose resort attracting New Yorkers until 1892, when it was bought by the State of New York for use as a quarantine facility to isolate suspected cholera carriers. When the cholera epidemic subsided, the facility was turned into a state park in 1908.[13]

The evolution of fast rail service connecting with passenger ferry boats to Fire Island, and the improvement of automobile highways out of the city by the 1920s, enabled upper middle class New Yorkers to vacation and weekend at Fire Island on a regular basis.

Today, 17 communities occupy segments of Fire Island between the ocean and

Great South Bay, some contiguous to each other, and others in relative isolation. Altogether, the developed areas comprise about one-fifth of the island's length. The remaining four-fifths of the island, including most of its eastern half, remains unoccupied (Figure 6-4). The entire island today, except for Robert Moses State Park at its western end, is within the boundaries of the Fire Island National Seashore. The National Park Service directly owns and manages 6235 acres of land and water, about one-third of the area within the total park boundary drawn by Congress. NPS has limited jurisdiction to influence land use concurrently with the preexisting summer communities within the remainder of the designated park area.

The summer communities of Fire Island have retained their treasured individuality and limited size due to the absence of any paved road extending the length of the island. Bridges provide access to public parks at each end but the intervening 30 miles of Fire Island is roadless, other than the beach and a few local service roads. Each community is connected to the mainland by passenger ferry service; residents leave their cars on Long Island and walk or bike on the island. This has fostered highly diverse social and cultural environments among the communities of Fire Island, ranging from artists and writers to investment

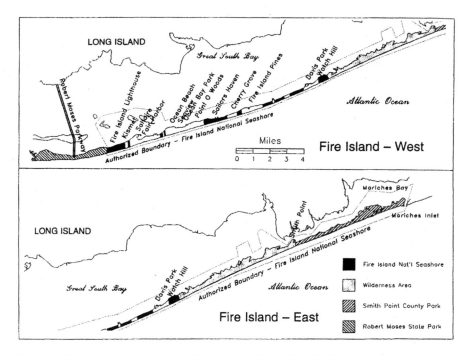

FIGURE 6-4 Summer communities and public parks of Fire Island. (Source: Matrix Maps)

bankers and lawyers. Certain communities like Fire Island Pines and Cherry Grove for decades have been vacation meccas for the affluent New York gay community. Ocean Beach and Ocean Bay Park cater to groups of young singles who rent houses together (known as "groupers"). Point O'Woods and Saltaire are family-centered and staidly conservative, where classic brown shingle cottages have belonged to the same family for several generations. The rest of the communities fill other niches in the socioeconomic web of greater New York, although all are costly locales in which to buy or rent a vacation cottage. Since one cannot drive on the island, local residents walk or cycle within their immediate communities. The rest of the public may drive to the state and county parks at either end of the island, or take a passenger ferry to the National Seashore units. Once on the beach, anyone can legally walk along the ocean as far as they desire.

The Legacy of Robert Moses

Despite the potential convenience of road access, Fire Island residents treasure their isolation from each other and from the crowding that a road would bring. For decades, they successfully resisted a crusade by Robert Moses, the "power broker" of New York State public works between the 1920s and the 1960s,[14] to build a road down the middle of the island. Moses was the dominant force in the planning and construction of public works on Long Island and in New York City from the 1920s until the 1960s. His populist ideals reflected in his early parks programs contrasted with the crude political expedients used to achieve them. His biographer, Robert Caro, drew a stark contrast between the earlier and later periods of Moses's career.

> In the beginning—and for decades of his career—the power Robert Moses amassed was the servant of his dreams, amassed for their sake, so that his gigantic city-shaping visions could become reality. . . . But little by little there came a change. Slowly but inexorably, he began to seek power for its own sake. . . . Increasingly, the projects became not ends but means—the means of obtaining more and more power.[15]

During the 1920s, as the Long Island State Park Commissioner, Moses launched an immense program of park and parkway construction, many of them laid out on the former estates of industrial tycoons. His crowning achievement was Jones Beach State Park, which serves millions of urban visitors every summer. From the 1920s on, he advocated an ocean parkway connecting Staten Island in New York City to Montauk at the eastern tip of Long Island that would traverse the length of each south shore barrier beach, including Fire Island. Moses proposed a Fire Island road incessantly between the late 1920s and the early 1960s. He triumphed in getting the Oceanview Highway built along the shore from Jones Beach as far east as Fire Island Inlet. He then connected the mainland to Robert Moses State Park on the western end of Fire Island via the Robert Moses Causeway. But there the project stopped, as Fire Islanders successfully prevented a road from continuing down the length of their island.

Hostility to the road and the development that would follow took a positive direction in the 1950s when islanders and conservationists rallied to save the Sunken Forest, a tract of semi-tropical habitat slightly below sea level behind the oceanfront dunes near the midpoint of the island. Two interests—preservation of a unique biome and hostility to a road on the island—were united by this enterprise.

In March 1962, over 100 Fire Island homes were destroyed by a devastating northeaster, the "Ash Wednesday Storm." In reaction, a state-sponsored committee was formed to make suggestions on storm hazard mitigation. Robert Moses, a member of the committee, crafted a proposal that would make his proposed road (now increased to four lanes) a structural protection measure against storms. It was approved. But in the face of determined local opposition, Governor Nelson Rockefeller asked for Moses to resign, thus quieting proponents of the road. Support for the road was further weakened when research showed that other barrier island roads had been undermined by erosion.

In June 1963 legislation was introduced to create the Fire Island National Seashore (FINS), following the model of the Cape Cod National Seashore established the same year. President Johnson signed Public Law 88-587 on September 11, 1964, establishing the FINS. This new park, created partly in reaction to the road building plans of Robert Moses, preserves sizeable tracts of undisturbed coastal barrier extending from the ocean to Great South Bay.

COASTAL HAZARDS OF FIRE ISLAND

Coastal hazards chronically threaten Fire Island. The island's east-west trending coastline and low elevation (generally a few feet above sea level) is vulnerable to storms tracking northward along the warm Gulf Stream. This vulnerable location, combined with the fragile nature of its stock of summer housing, places Fire Island at substantial risk from coastal storms, both hurricanes and winter northeasters. While risk to property is constant, the population at risk varies seasonally. There are only about 400 year-round residents, as compared with the average summer population of between 30,000 and 40,000.

The beachfront of Fire Island is largely shaped by nature. No seawalls have been constructed and sand-trapping jetties and groins are found only at the two inlets marking the east and west ends of the island, plus two small groins in front of the island's most populous community, the Village of Ocean Beach. Fire Island homeowners blame much of their ongoing erosion problem on jetties at

TABLE 6-1 Major Storms Striking Fire Island, N.Y., and Post-Disaster
Reports: 1938 to 1993

Storm Date	Storm Type	Major Post-Disaster Reports
September 1938	Hurricane	
September 1944	Hurricane	Joint Legislative Committee Studying Erosion on the Atlantic Shoreline (for 1938 and 1944 hurricanes)
November 1950	Northeaster	
November 1953	Northeaster	
August 1954	Hurricane Carol	
September 1960	Hurricane Donna	
March 1962	Northeaster	Temporary State Commission for Protection and Preservation of the Atlantic Shore Front
August 1976	Hurricane Belle	
February 1978	Northeaster	
March 1984	Northeaster	LIRPB: Hurricane Damage Mitigation Plan for the South Shore—Nassau and Suffolk Counties, N.Y.
September 1985	Hurricane Gloria	LIRPB: Proposed Long Island South Shore Hazard Management Program, 1989
August 1991	Hurricane Bob	Now and for the Future—A Vision of New York's Coast: Governor's Task Force on Coastal Resources
October 1991	Northeaster	
December 1992	Northeaster	Federal Emergency Management Agency 974 (Interagency Hazard Mitigation Team Report)
March 1993	Blizzard	1994 Report of the Governor's Coastal Erosion Task Force (winter storms of 1992/1993)

Source: Adapted from Governor's Coastal Erosion Task Force, 1994, 46–53.

Moriches Inlet at the eastward (updrift) end of the island, as well as on an incomplete groinfield built 30 years ago by the Corps of Engineers at Westhampton Beach, slightly farther east of the Moriches Inlet. Whether these projects trap enough sand to affect erosion rates some 20 miles to the west along Fire Island is currently in dispute.

Dramatic shifts in shoreline position are associated with, but not limited to, major storm events. In the past 60 years, Fire Island has experienced three major disasters: the Great Hurricane of 1938, the March 1962 Northeaster, and the series of northeasters during the winter of 1992/1993. These have been interspersed with numerous smaller storms and near misses (Table 6-1). These three major storms represent "threshold events." The physical and policy impacts of the Hurricane of 1938, the "Ash Wednesday" Storm of 1962, and the December 1992 northeaster, declared a "major disaster" by President Bush, have influenced public response to coastal erosion hazards both at Fire Island and at the national level.

On Long Island and its coastal barriers, the fast-moving Hurricane of 1938 brought winds over 100 mph, tidal inundation, and heavy rainfall. It caused at least 45 deaths on Long Island and its barrier beaches. From Jones Inlet to Montauk Point, the cost of damage to private structures, utilities, and government facilities along the south shore of Long Island totaled $6 million (1938 dollars).[16] On Fire Island alone it destroyed 265 houses and damaged many others.[17]Moriches and Shinnecock Inlets were respectively enlarged and formed by this storm.

Since the federal government was not yet involved in disaster assistance or flood insurance, the immediate response and recovery costs were shouldered by the property owners and charitable organizations. At this time federal interest was limited to technical studies conducted by the Corps of Engineers Beach Erosion Board established in 1930 at the urging of the American Shore and Beach Preservation Association.[18] In 1936 Congress adopted PL 74-834, which authorized federal assistance for constructing—but not maintaining—coastal protection works where "federal interests" were involved. After the Hurricane of 1944 struck south Florida, PL 79-166 authorized broader research on coastal erosion and substituted "public interest" for "federal interest." Federal cost-sharing in coastal erosion protection projects protecting public property was initiated at one-third of the total costs.[19]

After a series of hurricanes struck the Middle Atlantic states between 1954 and 1960, the Rivers and Harbors Act of 1960 authorized a Combined Beach Erosion Control and Hurricane Protection Project as a long-term solution to mitigate the effects of storms and storm-induced beach erosion along 83 miles of shoreline extending from Fire Island Inlet to Montauk Point. Today, that project is still authorized but much of it has not never been constructed.

The "Ash Wednesday Northeaster" of March 1962 prompted further expansion of the federal role in shore protection. The federal cost share in coastal projects was increased to 50 percent (PL 87-874) and 100 percent for coastal erosion studies. The Temporary New York State Commission for Protection and Preservation of the Atlantic Shore Front urged additional structural projects including seawalls, jetties, groins, and beach nourishment.[20] In response to this and other requests for protection, the North Atlantic Division of the Army Corps of Engineers undertook "Operation Five-High" as part of an emergency beach restoration project pumping 11.8 million cubic yards of sand to construct 85 miles of beaches from New York to Virginia. Seven emergency projects were undertaken in New York State, including Fire Island. "From Moriches Inlet to Fire Island Inlet, in the towns of Brookhaven and Islip, about 1.1 million cubic yards of material were placed along 8.5 miles of shore front to a top elevation of 12 feet above MLW. The project restored 9500 feet of dune and 37,000 feet of beach using sand dredged from various places in the Great South Bay at a total cost of $528,600."[21]

In 1964, a survey by geographers Robert Kates and Ian Burton found "a strong public demand for protective coastal engineering by governmental agencies."[22] They warned: "The United States is in danger today of embarking on a large-scale and costly program of coastal defense against storm hazard comparable with the program of flood control that has been operating in river valleys for the past quarter century."[23]

By the 1990s, however, the Fire Island beach and dunes were again substantially eroded, especially in front of most of the summer communities. (In natural areas, the dunes were able to migrate landward in the absence of houses.) In 1991, Hurricane Bob and the "Halloween Storm" caused extensive scouring of beaches and dunes. A year later, the northeaster of December 10–14, 1992, lashed the island with hurricane force winds, waves reaching 20 feet, and a series of high tides 4 to 6 feet above normal. At least 10 overwashes occurred along Fire Island, and two breaches opened at Westhampton Beach just east of the island. The destructive force of storm waves was particularly severe at points along the shoreline where the offshore bar was penetrated by deepwater gaps formed by rip currents.[24] This storm and the blizzard of March 1993 caused the destruction of over 35 homes on Fire Island and left dozens of others stranded on the narrow beachface, in various stages of collapse.

The winter of 1992/1993 left Fire Island devastated physically and financially. Even with federal assistance (discussed later), most homeowners experienced substantial out-of-pocket expenses to repair their homes and to share in the restoration of community facilities such as boardwalks and dune walkovers. Five communities, through beach erosion districts, funded local beach nourishment as stopgaps pending a larger-scale federal/state/local replenishment project proposed by the Army Corps of Engineers.

The great unknown was how public laws and policies would affect the rebuilding of damaged structures along the oceanfront. Sharing the responsibility for managing post-disaster actions were a galaxy of federal, state, and local units of government, each with its own priorities, powers, and politics. After reviewing the political geography of Fire Island, and the expert advice provided by post-disaster commissions, the outcome will be examined.

THE POLITICAL MOSAIC OF FIRE ISLAND

As the oceanfront of Fire Island retreats, dozens of homes are in the path of imminent destruction and hundreds more face collapse in the next two or three decades (Figures 6-5 and 6-6). Many have already been lost or relocated; others

FIGURE 6-5 (*Top*) A survivor of the old oceanfront row, as of 1980, now part of the dune line! (Photo by R.H. Platt) (*Bottom*) The demise of the same house after the 1992 Nor'easter. (Photo by Diane Abell, Fire Island National Seashore, National Park Service)

FIGURE 6-6 (*Top*) Before: oceanfront homes at Point O'Woods (older houses) and Ocean Bay Park (new house), 1980. The middle house had been moved landward by its owner. (Photo by R.H. Platt) (*Bottom*) After: Debris from collapse of house on right of Figure 6-5 (*Top*) after 1992 Nor'easter. (Photo by Alexander D. Platt)

179

despite severe storm damage have been rebuilt and even enlarged on the same site. Since the nation has assumed some of the costs of property damage through flood insurance and disaster assistance, the rebuilding and expansion of property at risk from coastal hazards at Fire Island and elsewhere is a national public policy concern. The larger society expects that the applicable governmental institutions will seek to reduce vulnerability to future coastal storm losses through the building and rebuilding process. But one looks in vain to identify a comprehensive, effective mitigation initiative at any level of government. Instead, one encounters a patchwork of public and private authorities with little zeal for reducing vulnerability to coastal hazards. Although a number of legal tools are available to promote mitigation at Fire Island, our review found that few are in fact utilized, and the market largely prevails.

Political and legal jurisdiction over Fire Island is divided both vertically and horizontally. In the vertical dimension there is a "legal stratigraphy" of federal, state, county, local, and private authorities overlaying the land area of Fire Island. In the horizontal dimension, Fire Island is an elongated mosaic of adjoining private and public units of jurisdiction. Response to the hazard of storms and shoreline erosion, including the recovery of developed shorelines after damaging events, is thus the product of a highly complex and fragmented array of authorities that often work at cross-purposes or simply pass the buck to each other.

Federal

It is an axiom of the American federal system that the federal government does not directly plan and manage local land use. However, the various coastal functions of federal agencies—such as national parks, flood insurance, disaster response, beach restoration—inject the federal government into local decision making in complex ways. Federal agencies involved with Fire Island include:

- National Park Service (NPS)—Fire Island National Seashore
- U.S. Fish and Wildlife Service (FWS)—Endangered Species
- Federal Emergency Management Agency (FEMA)—flood insurance and disaster assistance
- U.S. Army Corps of Engineers (COE)—inlet stabilization, beach nourishment, wetland permitting
- Office of Ocean and Coastal Resource Management of the National Oceanic and Atmospheric Administration (OCRM/NOAA)—coastal zone planning
- U.S. Environmental Protection Agency (EPA)—water quality, wetlands protection
- U.S. Coast Guard—navigation and marine safety

The *Fire Island National Seashore* (FINS) was established as a unit of the National Park System in 1964 by act of Congress,[25] which declared:

> For the purpose of conserving and preserving for the use of future generations certain relatively unspoiled and undeveloped beaches, dunes, and other natural features within Suffolk County, New York, which possess high values to the Nation as examples of unspoiled areas of great natural beauty in close proximity to large concentrations of urban population, the Secretary of the Interior is authorized to establish . . . the "Fire Island National Seashore" (PL 88-587, Sec. 1).

The national seashore boundary encompasses all of Fire Island east of Robert Moses State Park and all waters to a horizontal distance of 1000 feet in the Atlantic Ocean and up to 4000 feet in Great South Bay, including any islands and wetlands included therein. The total area including submerged lands within the FINS boundary amounts to 19,578 acres, of which the National Park Service (NPS) owns and manages 6242 acres.[26] The 17 Fire Island communities substantially retain their preexisting legal status. Conflicts abound between protecting biotic and other natural resources within national seashore holdings and tolerating private development and redevelopment within established communities. In 1980, Congress designated a 7-mile segment of the island, comprising some 1400 acres, as a federal wilderness area, the only such area in New York State. No homes are located within that area and park facilities are minimal.

The 1964 act granted authority to NPS to establish standards for local zoning in privately owned portions of the Seashore. Accordingly, NPS has divided the entire FINS into three districts: (1) "Community Development District," encompassing the 17 Fire Island communities; (2) "Seashore District," which includes all the rest of the FINS outside the community development district; and (3) "Dune District," which extends from mean high water line to 40 feet landward of the primary natural high dune crest.[27] New residential construction in the Community Development District is limited to single-family homes. Nonresidential construction must be reviewed by the local authority and the FINS Superintendent. Nonconforming uses may continue but may be modified only to become conforming. All new construction is prohibited within the Seashore District, including private lands outside of the 17 communities. In the Dune District, all uses are prohibited except approved vehicular and pedestrian crossings, soft-engineered dune stabilization measures, and scientific study. The Dune District was mapped in 1976 and now lies largely seaward of the present shoreline. It has not been updated and therefore is moot until its baseline is redrawn and adopted by Congress.

The FINS planning staff is authorized to review proposed zoning changes and to condemn property that does not comply with applicable zoning within the park boundaries. The superintendent may issue variances, exceptions, permits for commercial or industrial use, and special zoning permits.

human structures and the receding shoreline. They also advised against public investment in structural mitigation measures such as beach nourishment and dune building, and suggested that private cost-bearing be used instead.

By contrast, the Governor's Task Force on Coastal Erosion subsequent to the 1992/1993 northeasters dropped the concept of "retreat" and called for revival of the federal beach nourishment project originally authorized in 1960.

The Long Island Regional Planning Board in its 1984 *Hurricane Damage Mitigation Plan* reviewed the causes of persistent storm damage to New York's coastline. It reviewed erosion and flood control measures, land use and development patterns, environmental regulations, coordination of federal policies and evacuation, warning, and public education.[43] For Fire Island, it urged a more limited role for government and greater loss-bearing by the owners of affected summer homes.

(a)   "Government agencies should not provide funding for erosion control projects; . . . to protect seasonal homes . . . large-scale dune building and beach maintenance projects [are] not recommended."

(b)   "Individual owners should bear the brunt of structure loss due to erosion."

(c)   There is a need to "institute a uniform, dynamic dune district line for both the Towns of Islip and Brookhaven to protect the dunes."

(d)   The towns should "prohibit development [or] redevelopment within a uniform, dynamic dune district."

Five years later, the Board reinforced its "tough love" position in its 1989 *Proposed Long Island South Shore Hazard Management Program,*[44] whose goal was "to protect coastal resources and public amenities from impacts associated with shoreline erosion and flooding." The most radical feature of the plan was a recommendation to "phase out occupancy of Fire Island in the event that dwelling units are destroyed as a result of erosion and/or severe storms."[45] In addition, it suggested that "hard-structural" coastal erosion mitigation measures should be regulated by government agencies based on a better understanding of coastal processes and shoreline responses.

After Hurricane Bob and the Halloween Storm struck Long Island in 1991, New York Governor Mario Cuomo appointed a Task Force on Coastal Resources to address a variety of issues, including coastal storm damage, degradation of environmental quality, lack of public information, and coastal economic decline including shipping and fisheries. Its report deplored New York State's failure to realize its own coastal zone management goals: "While numerous federal, state, and local programs have been designed to discourage inappropriate use and development of coastal hazard areas, these efforts remain largely inconsequential compared to the problems and the complexity of government and private sector activities that affect the use and development of such areas."[46] It proposed that

" . . . an explicit policy of strategic retreat . . ." be developed so natural protective feature areas could serve to protect the coastline.[47] Furthermore, to prevent the destruction of the beach-dune system, it urged enactment or revision of regulatory setbacks and the strengthening of development regulations.

Three years later, a new Governor's Task Force on Coastal Erosion (GTFCE), appointed after the 1992/1993 storms, took a very different tack from its predecessor as well as the two reports of the LIRPB. In a two-volume report issued in September 1994[48] this task force, which included the president of the Fire Island Association among its members, focused on measures to hold the shoreline in place and protect existing property through sand bypassing, beach nourishment, and other measures (Table 6-2). While limited private cost-bearing of beach nourishment projects had been adopted by some Fire Island communities, retreat from the shoreline through setback districts and other regulatory mech-

TABLE 6-2 1993 Governor's Coastal Erosion Task Force Report: Priority Long-Term Actions Requiring Funding

- Bypass sand dredged from navigation channels in inlets to adjacent beaches.
- Initiate a program of routine sand bypassing at inlets and other constructed obstructions.
- Begin studies to design a sand bypassing program for inlets along the south shore of New York City and Long Island.
- Along the heavily developed Atlantic and Long Island Sound shorelines of New York City . . . hold the shoreline and provide a minimum 100-year storm protection.
- Along public recreation lands, maintain the beach to assure protection of recreational benefits to the public.
- Along the entire marine shoreline of the State, maintain the integrity of barrier landforms and natural beaches.
- For all capital projects, . . . the State cost share should be based on the value of project benefits which accrue to the State. [Until that is determined] . . . the present cost sharing arrangement should not be modified.
- Initiate a program of systematic monitoring of shoreline locations threatened by erosion.
- A monitoring program should be initiated with the goal of developing flood protection strategies.
- Support adequate research to ensure better decision making at the State, local, and individual levels.
- Local Waterfront Revitalization Programs should be used in developing comprehensive local erosion management programs.
- Develop a public awareness program on coastal hazards. . . .
- The reformulation phase of the Federal Fire Island Inlet to Montauk Point Beach Erosion Control and Hurricane Protection Project should be completed as quickly as possible.

Source: New York Governor's Coastal Erosion Task Force Report, 1994, Vol. II, 164–171

anisms had not been pursued. It proposed a twofold approach to deal with chronic flooding and erosion: (1) specific mitigation projects totaling $400 million in federal, state, and local funding and (2) improvements in emergency response. High priority projects included the elevation or protection of key evacuation routes, sand bypassing at inlets and groins, beach nourishment, and the filling of barrier island washover fans, as well as the creation of a reserve fund for rapid response to coastal hazards (such as sand placement after a storm). It also urged the Army Corps of Engineers to accelerate the reformulation of plans for the 30-year old Fire Island to Montauk Point Beach Erosion Control and Hurricane Protection Project.

Conspicuously, the 1994 report minimizes the role of land use regulation along eroding shorelines, with reference to the 1992 U.S. Supreme Court decision in *Lucas v. South Carolina Coastal Council*.[49] (See discussion in Chapter 5.) While the report frowns on further development in coastal hazard areas, it asserts that existing structures damaged or threatened by erosion should be acquired by government.

> Development should not occur in coastal hazard areas, i.e. V-zones and CEHAs. . . . Where appropriate, municipalities should improve their zoning regulations to provide for new development on larger lots with greater setbacks in such areas. However, increased area requirements will not resolve the problems of existing small lots or existing residences with limited setback. *Government acquisition of such properties may be the only viable alternative*[50] (emphasis added).

In other the words, the cost of losing a summer home to coastal erosion, beyond the value of flood insurance payment to the owner, should be borne by government, according to the task force. However, it did not propose state outlays specifically for that purpose.

The 1994 report called for the state to adopt Coastal Erosion Hazard Maps for Fire Island, with regulations to "be implemented by the local governments or the State."[51] But again it implies that such regulations cannot be enforced unless property owners are compensated: "State regulations and requirements for consistency with State and local coastal policies should be strengthened to ensure that new development is substantially limited in coastal hazard locations (V-zones and CEHA zones). This will help reduce State and local costs for disaster response. *Government acquisition of property may ultimately be required to prevent new development.*"[52] (Emphasis added.)

The series of New York State post-disaster reports summarized above share a common concern that coastal erosion is a costly and ongoing hazard to the South Shore of Long Island, including Fire Island. They differ in their approach and philosophy, however. In contrast to the emphasis on large-scale beach stabilization proposals in the 1960s, the 1984 and 1989 Long Island Regional Planning Board reports and the 1991 Governor's Task Force Report emphasized "retreat" and reduction of public subsidy to the protection of private property in the path of erosion. After the 1992/1993 winter storms and the *Lucas* decision, the 1994 Governor's Coastal Erosion Task Report abandoned retreat and returned to a "hold the line" philosophy, focusing on sand bypassing and beach replenishment. In any event, each of the reports has been largely ignored. Summer homes continue to be rebuilt larger than ever, the shoreline continues to recede, and the federal government continues to rush into the breach.

FEMA TO THE RESCUE

The devastation of the winter of 1992/1993 highlights the costs associated with maintaining status quo life on Fire Island and the marine coast of New York. The Federal Emergency Management Agency reported that direct federal disaster assistance through July 1993 totaled $233.6 million for the declared counties (Suffolk, Nassau, Rockland, Westchester, and New York City). This included $78.8 million in Individual Assistance, $78.7 million in Federal Flood Insurance, and $42.5 million in Public Assistance. The Federal Highway Administration provided $1.2 million for emergency repair work to the Federal-Aid Highway System. In addition, the Army Corps of Engineers approved $178,000 in disaster response funds and $5.7 million to close the breach at Westhampton.[53]

Specific data on NFIP policies for Fire Island is not readily available for several reasons. First, NFIP data is organized by municipality (Islip and Brookhaven in this case) and is not broken out for the individual communities of Fire Island (except for Ocean Beach and Saltaire, which are incorporated villages) (Table 6-3). Second, data is organized by policyholder name and address and would require an exhaustive search. But this is nearly impossible due to the lack of consistency of addresses on Fire Island. Third, access to individual policy data is limited by the Privacy Act.

Table 6-3 summarizes community-level NFIP data covering Fire Island (plus a bit more elsewhere in Brookhaven and Islip). The data is cumulative in that "92" includes all activity under each heading since 1978. The difference between "92" and "94" represents activity during the period covering the storms of 1992/1993. The last column displays the percentage change in NFIP payments, which approximately tripled during that period (except for a much larger jump for Saltaire). While the dollar amounts are not huge, the potential liability of the program at Fire Island is indicated by the total of $1.2 billion in coverage in 1994 (including mainland Islip and Brookhaven), which represents a 14.6 percent increase in coverage over 1992. A repetition of the 1938 Hurricane, the

Mitigation Not Accomplished

Pursuant to the Disaster Declaration of December 21, 1991, FEMA established an Interagency Hazard Mitigation Team to recommend ways to reduce future vulnerability in the affected areas. The team's report strongly urged remapping of shorelines and hazard zones: "A brief review of the South Shore FIRM [sic] indicates that the *Fire Island areas are the most in need of revision.* These FIRMS include the towns of Brookhaven and Islip, and the Villages of Ocean Beach and Saltaire."[59] (Emphasis added.)

The primary NFIP regulatory vehicle for guiding local management of flood-prone coasts is the designation of "coastal high-hazard zones" (V-zones) along open ocean shores and certain portions of bays and Great Lakes coasts. A coastal high-hazard area is "an area of special flood hazard extending from offshore to the inland limit of a primary frontal dune along an open coast and any other area subject to high-velocity wave action from storms or seismic sources."[60] Dunes are designated as landward limits of V-zones only where the dune cross-section above the 100-year stillwater elevation and seaward of the dune crest exceeds 540 square feet.[61] V-zones have long been delineated on flood insurance rate maps (FIRMs) based on computer simulation of the reach of a three-foot breaking wave during a 100-year storm event.

Given the imposing term "coastal high-hazard areas" and the costly modeling that underlies their delineation, one might expect that this would indeed be a zone of "strict regulation" where, presumably, new and rebuilt construction would be prohibited. This unfortunately is not the case. FEMA allows new construction in V-zones as far seaward as "the reach of mean high tide,"[62] provided its lowest floor is elevated to or above base flood elevation. Other NFIP requirements for V-zones prohibit "manmade alteration of sand dunes . . . which would increase potential flood damage"[63] and the enclosure of space below elevated structures with nonbreakaway walls or for purposes other than parking, access, or storage. Mapped V-zones typically are fairly narrow and are bordered on the landward side by A-zones with lower elevations and fewer restrictions.

While regulations for building in V-zones are surprisingly permissive, they comprise the primary NFIP limits on building and rebuilding on open coast shorelines. By definition, the area mapped as V-zones is highly dynamic and subject to rapid shifting during storms. Also, dunes defining the landward boundary of V-zones may be eroded to the point that they no longer satisfy the minimum cross-section criterion. Remapping of V-zones to update both their geographical location and the minimum elevations for construction is obviously needed after a major storm. At Fire Island, however, FEMA failed to follow the advice of its own mitigation task force regarding the need for remapping. As of 1997, the flood hazard areas of Fire Island had not been remapped and the effective flood insurance rate maps covering Fire Island predated the 1992/1993

FIGURE 6-8 Moving away from the water's edge, ca. 1996. (Photo by Alexander Platt)

storms. The V-zones are now seaward of mean high water in many places and the maps are thus obsolete.

The hazard mitigation task force report also recommended that the state impose "a moratorium . . . on reconstruction or new construction in areas identified as 25 feet inward of a bluff crest, back dune, or toe or vegetation line as they currently exist until the CEHA construction setback line is reestablished. . . ."[64] But no time length was specified for the moratorium and the state paid no attention to it. The task force also proposed that structures severely damaged or in danger of imminent collapse either be purchased under Section 1362 of the National Flood Insurance Act or be demolished or relocated under the 1987 Upton-Jones Amendment. In either case, the cost would be borne by the NFIP or the taxpayer, not the individual owner. Ultimately, no structures were acquired under these programs. However, a few homes were individually relocated (Figure 6-8).

Public Assistance

Islip and Brookhaven each applied for public assistance grants pursuant to the presidential disaster declaration. For Fire Island alone, the towns received approximately $1.6 million ($1.2 million in federal matching funds) in public assistance for emergency sand replacement, to remove debris, and to rebuild walkways. Fire Island communities received funds to rebuild public facilities, including dune walkovers at Fair Harbor ($148,000), a marina at Atlantique ($968,000), and emergency sand placement throughout developed portions of the island ($336,000). Individual assistance was apparently not utilized for Fire Island since the homes affected were seasonal and unoccupied at the time of the storm.

FEMA's contribution toward beach nourishment was limited to a few sites found eligible for "emergency assistance," which is intended "to save lives, to protect public health and safety, and to protect improved property."[65] Aside from this limited assistance, FEMA is not authorized to replace sand on "an unimproved, natural beach."[66] When a beach is considered "engineered" (i.e., designed to a specified elevation, width, and slope and subject to periodic maintenance), FEMA may provide more extensive sand replacement under its public assistance authority, subject to a 25 percent nonfederal cost share.[67]

Several communities since have undertaken local beach nourishment programs under erosion control districts, and the island is promoting a $60 million Corps of Engineers "interim" beach nourishment project as discussed below. Thus, ironically, the fact that Fire Island's beach was "unimproved and natural" at the time of the 1992 disaster prompted local action to fund sand replacement in the absence of FEMA public assistance funding. That action, especially if followed by a Corps of Engineers project, may obligate FEMA (and the federal taxpayer) to fund much more extensive sand replacement in the event of a subsequent disaster declaration for Fire Island.

DEFENDING PROPERTY RIGHTS AT THE WATER'S EDGE

The Fire Island Association (FIA) and Save Our Seashores (SOS) have each lobbied intensively to win government benefits for Fire Island while opposing any limits on the freedom of private owners to build or rebuild along the oceanfront. Between 1992 and 1994, FIA fought proposed federal legislation that would have limited new flood insurance coverage and the right to rebuild in areas of imminent erosion hazard. Meanwhile, both FIA and SOS since the early 1990s have pleaded for federal and state funding for beach nourishment. Thus, island homeowners, through their advocacy organizations, have sought to shift the blame for, and the costs of, oceanfront property damage to federal taxpayers, while maintaining that they should be free to use their property as they wish. This discussion considers Fire Island's role in defeating proposed erosion management policies under the National Flood Insurance Program. The following section examines its ongoing quest for federal and state beach nourishment.

In 1988, the Federal Emergency Management Agency (FEMA) contracted with the National Research Council (NRC), a research arm of the National Academy of Sciences, to provide expert advice on "appropriate erosion management strategies, supporting data needs, and applicable methodologies to administer these strategies through the National Flood Insurance Program."[68] This study was prompted by the 1987 "Upton-Jones Amendment" to the NFIP, which empowered FEMA to fund the demolition or relocation of insured structures in imminent risk of collapse. (See appendix.) The ensuing NRC report issued in 1990 urged FEMA to confront the need for federal standards for coastal erosion hazard areas. FEMA was advised to use historical shoreline change data to delineate areas expected to be eroded within 10, 30, and 60 years. Within 10-year zones—"areas of imminent collapse"—FEMA should deny new NFIP coverage (with legislative approval if necessary) and require local communities to prohibit new development. Seaward of the 60-year line, only readily moveable structures of less than 5000 square feet should be permitted. Insurance rates should include an erosion surcharge that progressively increases as the risk of loss becomes more imminent over time.[69]

FEMA then worked with congressional staff to draft legislation amending the NFIP in light of the NRC study. The House of Representatives passed H.R. 1236, the National Flood Insurance, Mitigation, and Erosion Management Act of 1991, by a vote of 388 to 18. The Senate version, S. 1650, however, encountered determined opposition from property rights, building, and real estate interests who argued that limits on new development, rebuilding, or flood insurance within the 10-year erosion zone would hurt property values. (The bill did not affect existing structures or their insurance coverage, unless a structure was in fact substantially damaged and was located in the zone of imminent collapse.)

Leading the charge was the Fire Island Association and its president, Gerard Stoddard, who in prepared testimony to Congress stated that:

> Coastal property owners will receive no benefit from S. 1650, should it become law. The principal drawback for property owners in erosion zones is that this legislation would take away their property without compensation and with no public purpose being served than the creation of additional open space. Homes in the 10-year erosion zone must be relocated or demolished, on pain of loss of their flood insurance policies, even in the absence of demonstrated negative impact on the environment. The standards suggested in S. 1650 aggressively eliminate private homes in the erosion zones, once they have been mapped.[70]

In his oral testimony, Stoddard continued:

> Our concern is when the program [NFIP] comes to accomplish other objectives than were originally contemplated. By that, it seems to us that the program is almost being subverted and being

turned into a social planning government land use program primarily in the area of requiring the mitigation. And when you
really look into what "mitigation" is about, it is about the
removal of the first row of coastal homes. . . . An awful lot of concern has arisen about S. 1650 after it passed the House, because
people suddenly realized what was involved here. What is
involved is the systematic removal of thousands of family residences, as well as vacation homes in families for generations.[71]

Although S. 1650 died in committee, its chief sponsor, Senator John Kerry
(D-MA), continued to champion erosion management in several subsequent
bills. In a hearing on S. 2907 (that still contained the National Academy of Sciences recommended changes), Senator Kerry hotly disputed the contention that
mapping and limitation of insurance in critical erosion areas would hurt property values.

As for the argument that the federal government should institute
programs but not establish criteria for participation, this is at the
outer border of irrationality. It is just a "gimme, gimme, gimme"
mentality, divorced from any conception of individual responsibility. . . . Rutherford Platt, in a letter dated June 15, 1992 [to
Senator Kerry] writes that the potential that property values will
drop upon the removal of flood insurance simply implies that the
government has been artificially propping up land values. Surely
the Contract with America folks would be dismayed to learn that
the federal government is so deeply involved in economic manip-
ulation.[72]

The Fire Island Association forged powerful alliances against the NFIP erosion management bills. Senator Alphonse D'Amato (R-NY), the ranking
Republican member on the Senate Banking Committee, vowed to filibuster any
flood insurance legislation and to accept no compromise on the issue. According
to a 1993 memo of the Coastal States Organization: "the Fire Island Association
is expected to place strong pressure on the Senator to keep any sort of land use
component out of flood insurance legislation."[73] Senator D'Amato in fact continued to block Senate adoption of the legislation until the provisions derived
from the National Academy of Sciences report were finally deleted.

Another ally of the Fire Island Association in 1992 surprisingly was C.M.
(Bud) Schauerte, head of the Federal Insurance Administration, the unit of
FEMA responsible for the National Flood Insurance Program. Upon assuming
this position, Schauerte testified on September 25, 1991, in favor of S. 1650,
stating: "This program will enable us to identify erosion-prone communities and

work with them to determine the erosion setback requirements so that, based on that knowledge, they can discourage unwise development of these coastal properties where severe property damage is inevitable."[74]

Six months later, however, Schauerte had joined the campaign against erosion management legislation in opposition to his own agency and his former position. As reported in a front page article in *The New York Times:*

> To the dismay of staff members at the Federal Emergency Management Agency who helped write the measure, one of its sharpest critics is C.M. (Bud) Schauerte. . . . Mr. Schauerte was appointed to head the [Federal Insurance Administration] after the revisions were well under way and has called them "back door Federal land use control" and "despicable."[75]

With the demise of S. 1650 at the end of the 102nd Congress, the Fire Island Association claimed victory for blocking the measure to that point but lamented the loss of their allies in the Bush administration.

> Fire Islanders can take some share of the credit for defeating the ill-advised flood insurance "reform" bills of 1991/1992. But they'd best gird their loins for a new attack in 1993. That assault will be against a backdrop of a national [Clinton] administration rather more tolerant of incursions against property rights than that of President Bush.[76]

A few months later, the Fire Island Association claimed total credit for the legislative outcome: "H.R. 1236 passed the house in 1992 [sic] by an astounding 388 to 18 vote. It was only because Fire Islanders sounded the alarm that a monumentally poor piece of legislation was not also approved by the Senate."[77]

The destruction of over 100 shorefront homes on Fire Island and substantial beach erosion during the winter storms of 1992/1993 failed to temper the opposition of the Fire Island Association to NFIP amendments on erosion management. Nor did the disaster declaration by President Bush on December 21, 1992, and the payment of some $24 million in NFIP claims to homeowners at Fire Island and its vicinity prompt any apparent second thoughts about the wisdom of building and rebuilding on eroding coasts.

On the other hand, Fire Island seized on the U.S. Supreme Court decision in *Lucas v. South Carolina Coastal Council,* issued on June 29, 1992, as support for their position against land use controls in erosion hazard areas. In that case, the U.S. Supreme Court, without discussing coastal hazards at all, held that regulations that entirely destroy property values may be unconstitutional takings in violation of the Fifth Amendment, unless supported by state common law property or nuisance doctrines. (See discussion in Chapter 5.) On behalf of the Fire Island Association, Stoddard summarized the majority opinion in his own words, *not those of the court,* as follows:

> The U.S. Supreme Court said the state has to have a better reason for taking, or in this case devaluing, someone's property than the desire of the legislature to protect the coastline. If you need the property for a public purpose, you can condemn it and buy it. But you can't make it into coastal open space by preventing the owner from using it for a reasonable purpose.[78]

Elsewhere, Stoddard claimed that "the U.S. Supreme Court said it will require close scrutiny of land use regulations that devalue private property."[79] (The Supreme Court opinion in fact said no such thing.)

The National Flood Insurance Reform Act of 1994,[80] reflecting the influence of the Fire Island Association and Senator D'Amato, contained no limitations on building or insurance availability in erosion hazard areas. This retreat from the stronger language of Senator Kerry's earlier bills (S. 1650, S. 1405, and S. 2907) reflected a political compromise to obtain improvements elsewhere in the National Flood Insurance Program. These provisions, which the Fire Island Association supported, included increased compliance by lenders, the creation of a flood mitigation fund within the NFIP, and additional coverage for the increased costs of compliance with "land use and control measures" for repetitively damaged properties.

Section 577 on "Evaluation of Erosion Hazards" directs FEMA to investigate the economic impacts of erosion and of various measures proposed to respond to it. FEMA is required to:

- List all communities that are likely to be identified as having erosion hazard areas
- Estimate the amount of flood insurance claims that are attributable to erosion
- Assess the full economic impact of erosion on the National Flood Insurance Fund
- Determine the costs and benefits of expenditures . . . to complete mapping of erosion hazard areas

Furthermore, Section 577 also calls for evaluation of the costs and benefits of mapping and management of erosion hazard areas in relation to:

- Property values in communities with erosion hazards
- Community tax revenues
- Employment
- Future economic development

- Disaster assistance costs
- Flood insurance claims

This study, as a substitute for the National Academy provisions, may be viewed as a delaying tactic on the part of opponents to erosion management. It appears to reflect the premise that erosion hazards should not be mapped if so doing would adversely affect coastal property values and the local tax base. That premise, if generally applied, would deter government from publishing any scientific data on natural hazards that might be detrimental to property values. It also exalts private property values and local tax revenue over other costs of erosion such as federal and state disaster assistance, public and private insurance costs, unreimbursed personal costs and emotional anguish ensuing from misinformed locational and investment decisions.

BEACH NOURISHMENT

While fighting proposals to manage erosion areas under the National Flood Insurance Program, the Fire Island Association and its allies have long pursued federal and state funding for beach nourishment. This represents the current chapter of a long and fitful history of Corps of Engineers involvement with the beachfront at Fire Island. A summary of that history provides a context to the present controversy.

Background

For over 40 years, the Army Corps of Engineers has proposed to restore beaches and dunes along the south shore of Long Island. The Fire Island to Montauk Point, N.Y., Beach Erosion Control and Hurricane Protection Project[81] was authorized by the Rivers and Harbors Act of 1960 in response to the risk of Long Island's south shore barrier islands being breached. Originally, the plan involved the widening of beaches to a uniform 100 feet, the raising of dunes to a minimum of 20 feet, and the construction of 50 groins.

The South Shore project authorized in 1960 preceded the enactment of the National Environmental Protection Act of 1969 (NEPA), which required an "environmental impact statement" (EIS) for "any major federal action significantly affecting the quality of the human environment." Fifteen groins were constructed at Westhampton Beach by 1970, but thereafter work was suspended to prepare an EIS as required by NEPA. A draft EIS was prepared on the project in 1975, followed by a final EIS in 1978. In the latter year, congressional appropriations for the project were halted due to the potentially negative environmental impacts of the proposed groins. The Environmental Protection Agency, Department of the Interior, and the Department of Commerce found the EIS unacceptable. The EIS did not adequately study alternatives, nor did it sufficiently address concerns about the groins' interference with sediment transport and shoreline processes. The Council for Environmental Quality (CEQ), accordingly, asked the Corps to reformulate the project in June of 1978 to satisfy these agencies. CEQ, however, consented to allow the Corps to construct protective "interim" measures along critical areas of the barrier islands as long as the measures could be reversed, if necessary, upon completion of the reformulation study.[82]

"Interim projects" are considered short-term solutions pending the reformulation of the original 1960 project for the entire Long Island South Shore. The interim projects must be designed based on the concept that long-term solutions would be provided upon completion of the reformulation study and must meet the approval of federal, state, and other interested agencies. The New York Coastal Erosion Hazard Areas Act requires that interim projects must provide erosion control for 30 years. Interim projects must also be "reversible" if the results of the reformulation study deem it necessary. If appropriate, the short-term measures will be integrated into the larger project.

In 1997, after decades of inaction, the reformulation study is back on track. Pursuant to the 1992 presidential disaster declaration, the Corps of Engineers headquarters approved a reformulation project management plan presented by the New York District, finding that:

> Along the length of the shoreline from Fire Island to Montauk Point, the width and the height of the beaches has been continually diminishing, thereby reducing the level of protection to the barrier islands and mainland areas. The loss of beaches has also proved to be a threat to the ecology of the bay areas that are contained within the project area. Breaches in the barrier islands resulting from the storm activity from the period of 1991 through 1993 have also reduced the level of protection and posed substantial threats.[83]

As the plan stands now, beaches along developed shorelines will be widened, dunes will be raised and planted with grass, and groins will be constructed as needed. Alternatives that may be considered include: no action, property buyout, revetments with or without beach restoration, breakwaters with or without beach restoration, seawalls with or without beach restoration, groins with or without beach restoration, inlet management plans/sand by-passing, beach restoration alone, removal or modification of existing groins, jetties, and revised land use regulations.[84]

The Westhampton Dunes Project

Just east of Fire Island, the Corps of Engineers has recently completed a beach and dune "interim project" that has given rise to a major building boom on reclaimed land at the newly incorporated Village of West Hampton Dunes. This project resulted from a class action lawsuit filed on behalf of property owners who contended that the groin construction built in the late 1960s under the original project was responsible for severe erosion and property damage (Figure 6-9) and in particular for two inlets that opened in the 1992 northeaster. The owners of property on Dune Road, just west (downdrift) of the groins, commenced the suit against the United States, the State of New York, and Suffolk County in 1993. As part of the settlement, the owners of "damaged or destroyed structures [may] rebuild whenever access and utilities are restored" to their property by the interim plan.[85] Over 170 structures are expected to be rebuilt according to this provision.[86]

Under the settlement, the Corps has constructed a 2000-foot steel wall to close the breach and pumped 4.5 million cubic feet of sand to restore the beach and dunes. The long-term cost has been estimated at $60–$75 million, of which 70 percent will be federal and 30 percent shared by the state and Suffolk County. Since then, the new Village of West Hampton Dunes, incorporated in 1993, has experienced a building boom on land newly reclaimed by the Corps project. Some municipal officers of the village have directly and indirectly speculated in

FIGURE 6-9 A derelict hulk standing in the ocean at Westhampton, Long Island, 1982. This area, downdrift from the Westhampton groinfield, breached in the 1992/1993 storms and was filled by the Corps of Engineers and redeveloped. (Photo by Edward Belt)

TABLE 6-4 Estimated Cost of Fire Island Interim Beach Nourishment Project

Fire Island Project, 1998	Total Cost	Federal	State	County*
Initial	$59M	$38.6M	$13.5M	$7.3M
Maintenance (per 3 yrs)	$5.3M	$3.4M	$1.2M	$0.6M
30 yr Total	$112M	$73M	$26M	$13M

Source: U.S. Army Corps of Engineers (unpublished data)

*Local communities may provide half of the initial county share or about $4 million.

real estate ventures within the village, which "owes its very existence to the federal, state, and county governments" and settlement of the class action lawsuit favorably to the property owners.[87]

By renourishing the beach, the federal government assures itself perpetual and increasing expenses. According to a 1995 National Research Council report on beach nourishment: "The level of protection is not absolute because of significant uncertainties about the frequency of storm conditions that may compromise project performance. The level of protection can be reduced rapidly during a major storm and is also progressively diminished when a previously nourished beach is not maintained by subsequent renourishment."[88] In addition, NRC found that "The inducement of shore development that might result from a beach nourishment project or program is an important public policy issue."[89]

The Fire Island Interim Plan

Fire Island hopes to be the next segment of Long Island's South Shore to receive federal funds for an interim beach nourishment project. The proposed project there would extend about 12 miles along the developed shoreline between Kismet and Davis Park. The Fire Island National Seashore Wilderness Area in the eastern part of the island would not be nourished under present plans. The project would include dune raising to 18 feet in front of Fire Island communities and a uniform 90 feet of dry beach width. The initial costs would amount to about $59 million, to be shared at a federal/nonfederal ratio of 65/35. With three-year renourishment, the total cost of the project over 30 years is estimated to be $112 million (Table 6-4).

Inspired by their neighbors to the east at Westhampton Dunes, some 400 Fire Island homeowners filed suit against the Corps of Engineers in 1995.[90] The plaintiffs demanded damages of $200 million due to "man-made accelerated recession" caused by inlet stabilization and groin construction at Westhampton, Shinnecock Inlet, the eastern end of Reach 2, and Moriches Inlet, the eastern end of Fire Island. The plaintiffs claimed that the "inlets and groin field caused between one-half and two-thirds of the shoreline recession on Fire Island over the last four decades," causing damage and a decline in real estate values.[91]

The claim that Fire Island communities located between 13 and 25 miles downdrift (west) of Moriches Inlet (and even farther from the Westhampton groinfield) were being affected by erosion due primarily to those projects is scientifically doubtful. Even two advocates of beach nourishment, coastal engineer Thomas Campbell and Joseph Vietri of the New York District Office of the Corps of Engineers, have questioned this theory of the cause of erosion along Fire Island.

> The pattern of erosion on Fire Island is not characteristic of an updrift deficiency (except possibly immediately adjacent to Moriches Inlet). The erosion rate from Moriches Inlet to Davis Park, the eastern beaches on Fire Island, is actually less (not more) than Davis Park to Kismet (the western beaches) which is the exact opposite of what would be expected for erosion caused by updrift deficiencies.[92]

Erosion in fact predominates along the Fire Island shore at an average rate of 0.5 meter (1.5 feet) per year.[93] But the standard deviation is high, reflecting great variation in observed erosion rates over different time scales and from one site to another.[94] In September 1997, the Federal Court dismissed the Fire Island suit against the Corps of Engineers.

The political case for federal and state assistance for beach nourishment for Fire Island is based in part on the claim that a breach in Fire Island would cause widespread flooding and damage along the mainland shore of Great South Bay. Some beach nourishment proponents have resorted to doomsday scenarios to try to sell their project as in an op-ed column in the Long Island newspaper, *Newsday,* by an island homeowner: "Right now, if you live on Long Island, you're less than 250 yards from disaster. That 250 yards is the width of parts of Fire Island, the only barrier keeping the Atlantic Ocean from flowing into the Great South Bay and onto the South Shore"[95] In the same vein, Save Our Seashores, Inc., in a written statement to the Governor's Coastal Erosion Task Force, declared that "175,000 people in Suffolk County alone would drown were they not to evacuate to higher ground" in the case of a breach of Fire Island in a Category I or II hurricane.[96]

Prophets of the demise of Long Island recalled a 1987 breach at Chatham, Massachusetts, which caused rapid erosion along a segment of mainland shore and the loss of several houses. But a new breach is likely to be less damaging in the wider, shallower Great South Bay than occurred at Chatham.

> A steeply sloped shoreline allows ocean swell to arrive without being slowed or changed until the last possible minute, resulting

in waves (especially during winter storms) that abruptly rise up
and break violently on the shoreline. . . . In contrast, the shallow
gentle slope which is typical along the south shore of Long
Island's mainland tends to reduce wave energy before it reaches
the shore.[97]

Inlets in fact may result from the buildup of storm surge in a bay that needs
to escape to the ocean, thus *lowering* flood levels along the mainland according to
coastal geologist (and stalwart foe of the Corps of Engineers) Orrin Pilkey:

> Since rescuing a few shorefront homes cannot justify [at least $60
> million for beach nourishment], the Corps has come up with a
> creative rationale. Its engineers say that replenishing the beach
> would prevent future storms from carving a new inlet across the
> island, thus saving thousands of homes on the Long Island "main-
> land" from flooding and destruction. . . . [But] new inlets form
> when a big storm pushes water into a lagoon behind a barrier
> island. As the storm moves inland, its winds change direction and
> blow back toward the ocean, forcing the piled-up water to rush
> back across the island and excavating a new channel. Opening a
> new inlet is like popping a cork in a bottle: Water flows out
> quickly, toward the sea. Mainland buildings are spared further
> damage.[98]

Even if a breach occurs, prompt action to close it may be more cost-effective
than to nourish the entire beach. In any event, several potential breach sites,
including at least one within the Wilderness Area, would not be fortified by the
interim beach nourishment currently proposed to apply only to part of the
island. In the event of a new breach, the Corps intends to close it immediately
before it can widen significantly. Advance permitting and stockpiling of sand
and transport equipment is now in progress in some locations.

Beach nourishment is fraught with interagency conflict. The National Park
Service (NPS) in general opposes shoreline stabilization and in particular objects
to beach nourishment along the 7-mile beachfront designated by Congress as a
Wilderness Area. The Fish and Wildlife Service (FWS) is concerned about the
degradation of habitat for the piping plover, a federally listed endangered
species. This bird nests on natural beach-dune areas and is displaced by artificial
beachfill. Both FWS and NPS urge that an environmental impact statement
should be prepared for a beach nourishment project by the Corps of Engineers
under the National Environmental Policy Act. The Stonybrook, Long Island,
field office of the FWS argues that the interim project is one element of the even-
tual Fire Island to Montauk Point project and an EIS is needed for the entire pro-

ject, and specifically for each of its components as they are proposed for construction. The Corps, however, has declined to prepare a full EIS, offering only a more limited "environmental assessment" (not available for public inspection) leading to a "finding of no significant impact" on the basis that the interim project will be of temporary duration and will theoretically be "reversible."

Meanwhile, the state of New York requires that a project must have a design lifetime of at least 30 years (representing a "44-year storm") as a condition of state participation. That strengthens the position of those who call for an EIS, since a 30-year project is not "temporary" (although the project itself may prove to be!).

Although available for interagency viewing, the draft Fire Island interim plan was not released to the public. The Department of the Interior (DOI) favors a 5-year rather than a 30-year design life unless a more rigorous EIS is completed. But a 5-year project would not be cost-effective according to the Corps economic analysis for the interim project. More than 90 percent of the project benefits would be derived from protection of the mainland from flooding (as called into question by Orrin Pilkey in his op-ed column quoted earlier). Indeed, even a 30-year design project may not generate the mainland protection benefits claimed for it since a "44-year storm" might not breach the island even without the project.

If a federal–state beach nourishment project is constructed, a key issue is the extent to which new or rebuilt oceanfront homes will be allowed. As of August 1998, no enforceable setback laws are in place. The National Park Service last mapped its "dune district" in 1976 and the primary dune crest of that time has now largely vanished into the ocean. Updating the boundaries of the district to reflect present conditions requires congressional action.

The New York Department of Environmental Conservation (DEC), as a condition to contributing to a beach nourishment project, proposes to establish building setback lines 25 feet behind the landward toe of the dune line under its Coastal Erosion Hazard Area authority. But if the new dune line is adopted as the baseline, construction will be partially allowed on land filled by the project. The state fears that it may have to compensate landowners if they cannot build on their rehabilitated lots.[99]

But this approach would mean that renourishment and dune rebuilding would open up new building lots or portions of lots on the oceanfront, as is happening at the newly filled breach at Westhampton Beach (where 200 homes are permitted to be rebuilt under the settlement reached between the property owners and the federal government).[100] New construction has already occurred in the communities of Fire Island Pines and Fair Harbor, just behind new dunes constructed by local beach erosion control districts. A six-month-long investigation by the Long Island newspaper, *Newsday*, disclosed the possibility that "at least four publicly funded beachfronts projects . . . could allow building of

200 to 300 new homes on previously damaged coastline."[101] While no federal subsidy is involved in these local projects, the homes will qualify for coverage under the National Flood Insurance Program. And their presence, together with older oceanfront homes, reinforces demands for a federal/state nourishment project to take up where the local projects leave off, both geographically and financially.

CONCLUSION

Over the past 40 years, the nation has developed an elaborate edifice of laws, programs, and policies for managing coastal hazard areas such as Fire Island, New York. These include the Coastal Zone Management Act, the National Flood Insurance Program, the Stafford Act (disaster assistance), the Clean Water Act, the Endangered Species Act, and various acts pertaining to the Army Corps of Engineers. In the case of the Fire Island National Seashore, Congress has provided special authority for the National Park Service to designate a "dune district" to restrict new oceanfront construction, even within preexisting summer communities. The state of New York also has been approved to receive federal coastal zone management funding, based on a suite of laws such as the 1981 Coastal Erosion Hazard Areas Act and the Tidal Wetlands Act. Local governments are invited to develop Local Waterfront Revitalization Plans to manage their shorelines under state and federal guidance.

But this patchwork of good intentions has proved futile in achieving a safe and sensible balance of human presence and nature along the shore of Fire Island. Counteracting these laws is another array of incentives to coastal development, including federal and state tax benefits, affordable flood insurance, disaster assistance, and federal grants for sewer and water systems and roads (although not at Fire Island), and beach nourishment funding. In the 1982 Coastal Barrier Resources Act, Congress suspended federal development incentives, including flood insurance for certain undeveloped coastal barriers designated as the Coastal Barrier Resources System (as expanded in 1990). But federal incentives remained available to support development on all other coastal barriers, including Fire Island.[102] The result, predictably, has been a one-sided struggle between the competing goals of reducing vulnerability to coastal hazards and optimizing the economic value of coastal property to the private owner/investor.

To recapitulate some of the findings of this case study:

- The National Flood Insurance Program paid $24 million in claims to Fire Island property owners after the 1992/1993 northeasters. Yet FEMA ignored its own post-disaster hazard mitigation team report by failing to remap the Fire Island shoreline after those storms and has not insisted that the state implement its Coastal Erosion Hazard Act as a condition of federal disaster assistance for the Long Island South Shore.

- The Army Corps of Engineers is proposing to construct a 30-year "interim" beach nourishment project at Fire Island without preparing a full environmental impact statement under the National Environmental Policy Act.
- The National Seashore Dune District line stands where it was drawn in 1976—out at sea.
- New York State still has not established CEHA setback lines at Fire Island although, after more than a decade of procrastination, it held public hearings on the topic in 1997. (Only the state Tidal Wetlands Act limited the reconstruction of a few homes directly on the eroded beachfront after the 1992/1993 storms.)
- The concept of orderly retreat from eroding shorelines, as endorsed by the 1991 Governor's Task Force on Coastal Resources and the Long Island Regional Planning Commission, was rejected by the 1994 Governor's Coastal Erosion Task Force, in favor of Fire Island's position that erosion is the fault of the Corps of Engineers and the shoreline should be stabilized through public spending.
- The Fire Island Association meanwhile claims credit for defeating proposed amendments to the National Flood Insurance Program to limit rebuilding and new insurance coverage in areas of imminent erosion hazard.
- Suffolk County abandoned enforcement of its septic system regulations to facilitate rebuilding after the 1992/1993 winter storms.
- The towns of Islip and Brookhaven do not yet have state-approved Local Waterfront Revitalization Programs.
- The 1992 U.S. Supreme Court decision in *Lucas v. South Carolina Coastal Council* has been invoked by Fire Island to argue that land use restrictions on eroding shorelines are unconstitutional.
- Two decades of research on shoreline change at Fire Island that attributes most sand loss to natural processes has been ignored.
- New construction is underway at Fair Harbor and Fire Island Pines immediately behind locally funded short-term beach-dune replenishment projects.

At Fire Island, the edifice of coastal laws has proven to be little more than a house of cards—or a sand castle. Politics has prevailed over policy and rules established by legislative bodies and regulatory agencies are just so many grains of sand. Fire Island stands defiantly on a southward facing shore of the Atlantic Ocean in the path of northeasters and hurricanes. And like all low-lying coastal areas, it confronts a gradually rising sea level. But science, law, technology, and common sense all pale in comparison with the monetary and psychic rewards of owning a very large house at the edge of the ocean, with the rest of the country behind you.

APPENDIX TO CHAPTER 6: A BRIEF HISTORY
OF COASTAL EROSION AND THE NFIP

As originally adopted in 1968, the National Flood Insurance Act was silent on the subject of coastal erosion. The Flood Disaster Protection Act of 1973,[103] in response to recent erosion losses along the Great Lakes, amended the NFIP to include erosion and mudslides as insurable hazards.

> The term "flood" shall also include the collapse or subsidence of land along the shore of a lake or other body of water as a result of erosion or undermining caused by waves or currents of water exceeding anticipated cyclical levels, and all of the provisions of this title shall apply with respect to such collapse or subsidence in the same manner and to the same extent as with respect to floods . . . including the provisions relating to land management and use. . . .[104]

The Federal Insurance Administration in 1976 issued new regulations to reflect this mandate. First, "areas of special flood-related erosion hazards" were defined as a separate and distinct hazard area category, to be designated as "E-zones" on flood hazard maps.[105] Second, a statement of purpose was adopted "that all eligible communities must take into account flood, mudslide . . . and flood-related erosion hazards, to the extent they are known, in all official actions relating to land management and use."[106] This regulatory (and legislative) definition of erosion hazards is limited because it excludes erosion that occurs gradually unrelated to a flood event. Still it did establish a framework for communities to address the problem. If and when E-zones are delineated by the Federal Insurance Administration, the affected coastal community must "require setbacks for all new development from the ocean, lake, bay, riverfront, or other body of water to create a safety buffer consisting of a natural vegetative or contour strip."[107]

The Federal Insurance Administration, however, failed to implement these provisions. No E-zones have been designated on flood insurance rate maps (FIRMS), no "safety buffers" have been designated, and no mandatory community land use management measures for erosion hazards have been required. In July 1977, at a conference on coastal erosion convened by the Federal Insurance Administration, it was proposed that the administration should declare a "no insurance zone" in high-risk erosion areas: "This is the zone of imminent failure or imminent destruction. You know the building is going to drop over within the next 10 years, and so, its not an insurable situation."[108] That proposal, unadopted at the time, was to be a key provision of legislation opposed by the Fire Island Association 15 years later. The 1977 conference concluded with three recommendations:

1. The National Flood Insurance Program legislation should be changed to eliminate the need to distinguish between anticipated and unanticipated erosion
2. Land use regulation should be emphasized in any program addressing erosion hazards
3. Recession rate analysis methods must be evaluated further and any overall standards should accommodate state or regional differences

While the federal government continued to procrastinate in setting erosion management standards, several coastal states, starting with North Carolina in 1974, began to adopt their own erosion setback programs. New York State in 1981 adopted its Coastal Erosion Hazard Areas Act, which, however, has not been applied to the South Shore of Long Island, including Fire Island.

Federal creation of a "no insurance zone" appeared in a different form in the Coastal Barrier Resources Act of 1982, whose adoption by Congress was in response to research by H. Crane Miller and John R, Sheaffer on the effects of Hurricane Frederic (1979). The act prohibited new NFIP coverage and various other federal incentives to development for designated undeveloped and unprotected coastal barriers. As enlarged in 1990, the "Coastal Barrier Resource System" established by the act is the only instance of denial of flood insurance benefits to discourage (but not prohibit) development in designated areas. It does not apply to shorelines that are already developed.

Developed shores in the Southeast and the Great Lakes continued to experience widespread loss of structures due to erosion in the mid-1980s. Senator Walter Jones of North Carolina joined Rep. Fred Upton of Michigan in sponsoring what became known as the "Upton-Jones Amendment" to the National Flood Insurance Program in 1987. The amendment authorized the NFIP to offer two options to owners of insured structures faced with imminent risk of collapse due to erosion: (1) a payment of up to 40 percent of the insured value to relocate the structure to a safer site, or (2) a payment of up to 110 percent to demolish the structure and remove the debris. The "Upton-Jones" program only resulted in payments for about 300 structures, most of which were demolished, thus not saving any money to the NFIP. It was abolished in the 1994 NFIP amendments.

At this writing (summer 1998), FEMA is conducting a study of the benefits and costs of erosion hazard mapping pursuant to the compromise between Senators Kerry and D'Amato that resulted in Section 577 of the 1994 Flood Insurance Reform Act.

NOTES

1. C.E. Johnson (1982). *Historic and Geomorphic Evidence of Barrier Dynamics and the Origin of the Sunken Forest, South Shore of Long Island, New York.* M.S. thesis, University of Massachusetts at Amherst.

2. Larry R. McCormick and M.A. Toscano (1981). "Origin of the Barrier System of

Long Island, New York." *Northeastern Geology* 3(3/4): 230–234; Stephen P. Leatherman (1985). "Geomorphic and Stratigraphic Analysis of Fire Island, New York." *Marine Geology* 63(1985): 173–195.

3. Larry R. McCormick, Orrin H. Pilkey, Jr., William J. Neal, and Orrin H. Pilkey, Sr. (1984). *Living with Long Island's South Shore.* Durham, N.C.: Duke University Press.

4. Ibid., 58.

5. James R. Allen and Norbert P. Psuty (1987). "Morphodynamics of a single-barred beach with a rip channel, Fire Island, N.Y." *Coastal Sediments '87 Proceedings of the ASCE.* New Orleans: American Society of Coastal Engineers.

6. McCormick et al., note 3, 6.

7. Cashin Associates (1993). *The Environmental Impacts of Barrier Island Breaching with Particular Focus on the South Shore of Long Island, New York.* Hauppauge, N.Y.: Cashin Associates.

8. Ibid., 15.

9. M.L. Schwartz (1982). *The Encyclopedia of Beaches and Coastal Environments.* Stroudsburg, Penn.: Hutchinson Ross Publishing Co.

10. Paul J. Godfrey (1976). *Comparative Ecology of East Coast Barrier Islands: Hydrology, Soil, Vegetation.* Contribution No. 2 in Barrier Islands and Beaches Technical Proceedings of the 1976 Barrier Islands Workshop. Washington, D. C.: The Conservation Foundation.

11. Ibid.

12. M.C. Johnson (1983). *Fire Island 1650s–1980s.* Mountainside, N. J.: Shoreland Press.

13. Ibid.

14. Robert A. Caro (1974). *The Power Broker: Robert Moses and the Fall of New York.* New York: Knopf.

15. Ibid., 19.

16. N.Y. Governor's Coastal Erosion Task Force (1994). *Final Report,* Vol. II. Albany: New York Department of State, 45–46.

17. Thomas J. Campbell and Joseph Vietri (1994). "Storm Impacts and Response on the South Shore of Long Island: Fire Island Inlet to Montauk Point," in *Alternative Technologies in Beach Preservation.* Proceedings of the 1994 National Conference on Beach Preservation Technology. New York: American Society of Civil Engineers, 488–507.

18. Rutherford H. Platt et al. (1992). *Coastal Erosion: Has Retreat Sounded?* Program on Environment and Behavior Monograph No. 53. Boulder: Institute of Behavioral Science, University of Colorado, 24.

19. Ibid.

20. The 1962 report also urged that nonstructural measures (e.g., zoning regulations limiting shorefront construction and public land acquisition for coastal recreation) be employed. This acknowledgment of nonstructural mitigation measures was a breakthrough in the state but was not implemented.

21. N.Y. Governor's Task Force, note 16, 51.

22. Ian Burton, Robert W. Kates, and Rodney Snead (1969). *The Human Ecology of Coastal Flood Hazard in Megalopolis.* Research Paper No. 115. Chicago: University of Chicago Department of Geography, 1.

23. Ian Burton and Robert W. Kates (1964). "The Floodplain and the Seashore: A Comparative Analysis of Hazard-Zone Occupance." *Geographical Review* 54(2): 366–385, 366.

24. First Coastal Corp. (1993). *Fire Island Erosion and Damage Assessment: 1992 Nor'easter—December 10–14, 1992.* Westhampton Beach, N.Y.: (mimeo).

25. PL 88-587; 16 USCA sec. 459e et seq.

26. The boundary of the park includes approximately 15,700 acres of underwater lands and 650 acres on the Long Island mainland, the former William Floyd estate. This leaves only about 3150 acres of land area on Fire Island actually owned by the National Park Service. (Diane Abell, FINS planning office, pers. comm., June 15, 1998).

27. 36 CFR Part 28.3(d).

28. PL 92-583; 16 USCA sec. 1451 et seq., as amended.

29. Executive Law 42—The New York Waterfront Revitalization and Coastal Resources Act.

30. For a more detailed summary, see Platt et al., note 18, 81–92.

31. N.Y. Environmental Conservation Law, Art. 34.

32. N.Y. Governor's Coastal Erosion Task Force, note 16, 62.

33. William Daley, N.Y. Department of Environmental Conservation, pers. comm, September 5, 1997.

34. New York Department of Environmental Conservation (1992). *Tidal Wetlands Land Use Regulations 6 NYCRR Part 661.* Albany, NY DEC, 1.

35. Ibid. The act regulates excavation, dredging, construction, pollution, subdivisions, and other activities affecting wetlands. Permits are required for buildings, roads, pilings, and other encroachments in both wetlands and adjacent lands, the regulated areas of the act. Coastal wetlands are defined as those lands "from the water to the mean high water mark." Adjacent lands (upland) are measured from mean high water, upland for 300 feet, or to the 10-foot upland contour line, whichever is closer.

36. Ibid., 11. Over 200 Fire Island projects were reviewed by DEC under the authority of the Tidal Wetlands Act during 1993. The wetlands permit process resulted in denials for approximately 30 applications to rebuild decks, pools, and homes along Fire Island's oceanfront following the December 1992 northeaster.

37. For an approved subsurface sewage disposal system, homes must have a professional plan drawn up that requires a 75-foot setback from surface water sources, a 5 percent slope for 20 feet from the edge of the septic system, and a minimum of 1 foot of fill above the septic tank lid. M.H. Hibberd and J.H. Baier (1986). *Standards for Approval of Plans and Construction for Subsurface Sewage Disposal Systems for Single-Family Residences.* Riverhead, N.Y.: Suffolk County Department of Health Services, Division of Environmental Quality.

38. The board is a quasi-governmental agency established following guidelines set in the Planning and Zoning Laws of New York State (Article 12-B).

39. Long Island Regional Planning Board (1984). *Hurricane Damage Mitigation Plan for the South Shore—Nassau and Suffolk Counties, NY.* Hauppauge, N.Y.: LIRPB; Ibid. (1989). *Proposed Long Island South Shore Hazard Management Program.* Hauppauge, N.Y.: LIRPB.

40. David J. Scherf (1995). *Reconstruction of Shorefront Structures at Fair Harbor and Ocean Bay Park, Fire Island, New York: Case Studies of Land Use Control in the Dynamic Coastal*

Fringe. Unpublished Master of Science thesis, University of Massachusetts at Amherst Department of Geology and Geography.

41. Save Our Seashore (1993). "Statement to Governor's Task Force on Coastal Erosion Presented by Jane Ekstein, Chair Save Our Seashore Emergency Coalition May 5, 1993." Brightwaters, N.Y.: S.O.S. Emergency Coalition.

42. FEMA's 1993 Hazard Mitigation Report issued after the 1992 northeaster is discussed later in this chapter.

43. Lee E. Koppelman (1984). *Hurricane Damage Mitigation Plan for the South Shore— Nassau and Suffolk Counties, N.Y.* Hauppauge, N.Y.: Long Island Regional Planning Board.

44. Lee E. Koppelman (1989). *Proposed Long Island South Shore Hazard Management Program*. Hauppauge, N.Y.: Long Island Regional Planning Board.

45. Ibid., 3–23.

46. Governor's Task Force on Coastal Resources (1991). *Now and for the Future: A Vision for New York's Coast*. Albany: GTFCR, 55.

47. Ibid., 56.

48. N.Y. Governor's Coastal Erosion Task Force, note 16.

49. Ibid., Vol. 2, 81-82.

50. Ibid., Vol. 1, 11.

51. Ibid., Vol. 2, 178.

52. Ibid., 172.

53. Ibid., 5.

54. Thomas Maier and John Riley (1998). "A Shoreline in Peril" (five-part series). *Newsday* (Aug. 16–20).

55. John Riley (1998). "Flood of Claims." *Newsday* (Aug. 18), A8, A24–A25.

56. Ibid.

57. FEMA unpublished data. No state or local share was required for these proposed buy-outs, which were never in fact accomplished due to the repeal of Section 1362 by the Flood Insurance Reform Act of 1994.

58. Laurie Cohen (1997). "Getting a Beachfront Home, the Hard Way." *The Wall Street Journal* (Jan. 31), B10.

59. Interagency Hazard Mitigation Team (1992). "Report in Response to the December 21, 1992 Disaster Declaration for the State of New York." New York: Federal Emergency Management Agency (mimeo), 26.

60. 44 CFR sec. 59.1.

61. 44 CFR sec. 65.11.

62. 44 CFR sec. 60.3(e)(3).

63. 44 CFR sec. 60.3(e)(7).

64. Note 59, 56.

65. 44 CFR sec. 206.225.

66. 44 CFR sec. 206.226(h)(1).

67. 44 CFR sec. 206.226(h)(2).

68. National Research Council (1990). *Managing Coastal Erosion*. Washington: National Academy Press, 3.

69. Ibid., 5–9.

70. Hearing before the Subcommittee on Housing and Urban Affairs of the Senate Committee on Banking, Housing, and Urban Affairs. (102nd Cong, 1st Sess.), September 25, 1991. Washington, D.C.: USGPO.

71. Ibid.

72. Hearing on S. 2907 before the Subcommittee on Housing and Urban Affairs of the Senate Committee on Banking, Housing, and Urban Affairs. (102nd Cong, 2nd Sess.), July 27, 1992. Washington, D.C.: USGPO, 17.

73. Discussion paper by Kerry Kehoe, Coastal States Organization to Jeff Benoit and others re: "Flood Insurance Reform in the 103rd Congress," January 20, 1993.

74. Testimony on S. 1650 before the Housing and Urban Affairs Subcommittee of the Senate Committee on Banking, Housing, and Urban Affairs, September 25, 1991. Quoted in Letter to the Editor of *The New York Times* from Rutherford H. Platt, June 23, 1992.

75. Cornelia Dean (1992). "Beachfront Owners Face Possible Insurance Cuts," *The New York Times* (May 27), A1 and A16.

76. Gerard Stoddard (1993). "New Flood Bill Introduced: Once More Into the Breach," *Fire Island Association Newsletter* (May-June-July issue).

77. Gerard Stoddard (1993). "New Flood Insurance Bill—Much Improved," *Fire Island News* (Aug. 26–Sept. 1), 6.

78. Gerard Stoddard (1992). "Public Policy Issues in Coastal Management Initiatives." Remarks presented at the Interdisciplinary Hurricane Conference, Hofstra University (Nov. 6).

79. Gerard Stoddard (1995). "Coastal Policy Implications of Right to Rebuild Questions" *Shore and Beach* (Jan.): 25–35, at p. 30.

80. PL 103-325.

81. Originally set forth in House Document 425, 86th Congress, 2nd Session, 21 June 1960.

82. U.S. Army Corps of Engineers (1986). Unpublished Memorandum: "Fire Island Inlet to Montauk Point, N.Y.—Beach Erosion Control and Hurricane Protection." (Sept. 30).

83. N.Y. District Office, Army Corps of Engineers (1996). "Project Background and Description." Memo re: "Environmental Scoping Document for Storm Damage Reduction Reformulation Study. Atlantic Coast of Long Island, Fire Island Inlet to Montauk Point, NY."

84. Ibid.

85. U.S. Army Corps of Engineers (1995). *Fire Island to Montauk Point, N.Y. Moriches to Shinnecock Reach Interim Plan for Storm Damage Protection Technical Support Document.* New York: USACE.

86. Ibid.

87. Thomas Maier (1998). "It Took a Village," *Newsday.* August 19, A8, A46–A47.

88. National Research Council (1995). *Beach Nourishment and Protection.* Washington, D.C.: National Academy Press, 12.

89. Ibid., 3.

90. Homeowners were solicited to join the suit in a letter of March 10, 1995 from the "Fire Island Litigation Committee" distributed by the same New York City law firm that represented the Westhampton Beach plaintiffs.

91. Complaint filed February 9, 1996, in U.S. Court of Federal Claims by Thomas DeVito et al. against the United States of America.

92. Thomas J. Campbell and Joseph Vietri (1994). "Storm Impacts and Response of the South Shore of Long Island: Fire Island to Montauk Point," in *Alternative Technologies in Beach Preservation.* Proceedings of the 1994 National Conference on Beach Preservation Technology. New York: American Society of Civil Engineers, 488–507.

93. James R. Allen and Norbert P. Psuty (1987). "Morphodynamics of a Single-Barred Beach with a Rip Channel, Fire Island, NY," in *Coastal Sediments '87.* New Orleans: Waterworks Division/ASCE.

94. James R. Allen and C. LaBash, "Time and Space Scales of Shoreline Change: Living along the Edge at Fire Island National Seashore, New York" (in press).

95. David Fisher (1992). "Viewpoints: The Surf Crashes for Thee." *Newsday* (Dec. 22), 73.

96. Save Our Seashore, Inc. "Statement to the Governor's Task Force on Coastal Erosion," Oct. 6, 1993.

97. Cashin Associates (1993). *Scientific Literature Review on the Environmental Impacts of Barrier Island Breaching with a particular focus on the South Shore of Long Island, New York.* Hauppauge, N.Y., p. 13, citing Willard Bascom (1964), *Waves and Beaches.* New York: Doubleday.

98. Orrin Pilkey (1996). "Sand-Castle Science on Fire Island" (op-ed column). *The New York Times* (Aug. 18), 15.

99. Fred Anders, NY Dept. of State, Division of Coastal Resources and Waterfront Revitalization, Pers. Comm. Sept. 22, 1997.

100. Diane Ketcham (1997). "Despite Elements, Governments and a Bird, a Village is Born," *The New York Times* (Jan. 20), B1 and B5.

101. Thomas Maier and John Rilery (1998). "On New Dunes, Building Boom." *Newsday* (Aug. 16), A4, A5, and A32–A36.

102. Rutherford H. Platt, Sheila G. Pelczarski, and Barbara K. R. Burbank (1987). *Cities on the Beach: Management Issues of Developed Coastal Barriers.* Research Paper No. 224. Chicago: University of Chicago Department of Geography.

103. PL 93-234.

104. National Flood Insurance Act, Sec. 1370(c); codified at 42 USCA Sec. 4104.

105. 44 CFR sec. 65.1.

106. 44 CFR sec. 60.1.

107. 44 CFR sec. 60.5(b)(2).

108. John R. Sheaffer (1977). Comments in *Proceedings of the National Conference on Coastal Erosion.* Washington, D.C.: Federal Insurance Administration, 29.

CHAPTER 7

St. Charles County, Missouri: Federal Dollars and the 1993 Midwest Flood

Miriam Gradie Anderson and Rutherford H. Platt

Ten thousand river commissions, with the minds of the world at their back, cannot tame that lawless stream, cannot curb it or confine it, cannot say to it, "Go here," or "Go there," and make it obey; cannot save a shore which it has sentenced; cannot bar its path with an obstruction which it will not tear down, dance over, and laugh at.

Mark Twain, *Life on the Mississippi* (1896)

INTRODUCTION

The Mississippi/Missouri River system ranks third in the world in main stem length (behind the Amazon and the Nile) and also in drainage area (behind the Amazon and the Congo).[1] The 1.2-million square-mile Mississippi/Missouri river basin contains most of the major inland metropolitan regions of the United States including Minneapolis–St. Paul, St. Louis, Kansas City, Pittsburgh, Cincinnati, Memphis, Denver, New Orleans, and a large part of Chicago. The main stem rivers and their major tributaries have historically nurtured the economic growth of all these cities except Denver. Even before the modern era of "river improvements," the importance of the inland waterways as arteries for transporting the agricultural produce of the region was ecstatically described by Anthony Trollope during his travels there in 1861–1862.

No river on the face of the globe forms a highway for the produce of so wide an extent of agricultural land. The Mississippi with its

215

tributaries carried to market, before the [Civil] war, the produce of Wisconsin, Minnesota, Iowa, Illinois, Indiana, Ohio, Kentucky, Tennessee, Missouri, Kansas, Arkansas, Mississippi, and Louisiana. This country is larger than England, Ireland, Scotland, Holland, Belgium, France, Germany, and Spain together. . . . The states named comprise the great center valley of the continent, and are the farming lands and garden grounds of the western world.[2]

The Mississippi system drains a region of great climate variation. Average annual precipitation ranges from less than 20 inches in the semiarid grasslands of the upper Missouri Basin in the west to more than 50 inches in the Ohio River Basin in the east. High levels of river flow result from seasonal snowmelt in the northern portions of the basin and from storm systems that move across the region from west to east at any time of year. Tropical hurricanes emerging from the Gulf of Mexico in late summer and fall occasionally threaten the lower valley with heavy rainfall, wind, and coastal storm surge. All of these climatic phenomena contribute to recurrent and occasionally massive floods along the Mississippi and its tributaries. Since the earliest European settlement in the valley in the eighteenth century, Mississippi River floods have inflicted ever greater economic and social burdens on the populace and cities of the region and therefore on the nation. More than any other river system, Mississippi floods have consistently attracted national attention and, in times of disaster, national response. The policies of the United States toward flooding have to a great extent been forged in the incessant quest for a reasonable accommodation between human settlement and natural hazard in the Mississippi River valley.[3]

In the summer of 1993, the Upper Midwest—comprising 30 percent of this immense river basin—experienced some of the worst flooding in its recorded history.[4] After a wet fall and spring, the area from the northern plains into the central Corn Belt received up to 250 percent of its normal summer rainfall between June and August. The result was an onslaught of devastating floods

TABLE 7-1 Comparison of 1927 Lower Mississippi Valley Flood and 1993 Upper Mississippi Valley Flood

	1927 Flood	1993 Flood
Area Flooded	12.8 million acres	20 million acres
Economic and Property Damage	$12 billion ($1993)	$18 billion
Deaths	245–500	48
Buildings Damaged	137,000	45,000 (in floodplains)
People Made Homeless	700,000	70,000

Adapted from James M. Wright, "Transcending the 1993 Midwest Flood: National Policy and Program Implications." Published by the author, 1994, Table 1.

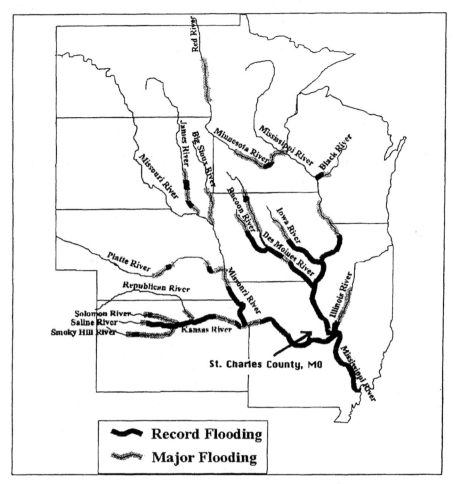

FIGURE 7-1 Areas of major flooding during the Midwest Floods of 1993 and the location of St. Charles County, Missouri. (Adapted from 1994 Report of the Interagency Floodplain Management Review Committee, Figure 1.6)

exceeding the 1-percent annual chance ("100-year flood") in many locations that inundated an estimated 6.5 million acres in nine states.[5] The flooding actually was more widespread than the landmark 1927 Lower Mississippi Valley Flood although the human toll was fortunately lower (Table 7-1, Figure 7-1).

The 1993 Midwest floods caused at least 48 deaths, $18 billion in overall damage (including secondary and tertiary impacts), and displaced tens of thousands of households and businesses.[6] Federal outlays related to the flood amounted to approximately $4.3 billion in direct expenditures, $1.3 billion in federal insurance payments ($297 million from the National Flood Insurance Program and $1.03 billion in federal crop insurance payments), and $0.6 billion in disaster loans to individuals, businesses, and communities.[7]

The high level of flood damage resulted in massive amounts of federal dollars being used to repair public facilities and infrastructure, rebuild or remove private homes, reimburse farmers, and assist individuals in reestablishing their lives. The amount of federal funds allocated for recovery and the ultimate use of those funds was unprecedented for the Midwest. The floods of 1993 coming so quickly on the heels of other substantial natural disasters, forced the federal government to reexamine its role in the recovery phase and to reconsider how federal dollars were being used by individuals and local governments. St. Charles County, because of its location, level of damage, and federal dollars received, became for many an example of both the best and the worst aspects of federal intervention in floodplain management.

ST. CHARLES COUNTY: GEOGRAPHY AND FLOOD HISTORY

St. Charles County, Missouri, occupies 586 square miles, equivalent to half the area of Rhode Island. Roughly wedge-shaped, the county is bounded on the north by the Mississippi and on the south by the Missouri with a total of 108 miles of river shoreline. About 43 percent of St. Charles County (252 square miles) is identified as floodplain, roughly one-third along local streams and the rest in the bottomlands adjoining the Mississippi and Missouri Rivers (Figure 7-2).

The county in fact lies in the path of two inexorable and irreconcilable forces: water and people. On the one hand, it is wedged between the Mississippi and the Missouri Rivers, which meet at the county's easternmost tip where "Old Man River" gets serious about rolling south toward the Gulf of Mexico. Above this confluence lies the Upper Mississippi River Basin and the entire Missouri River Basin. The runoff of nearly one-fourth of the land area of the conterminous United States, extending from Chicago to Montana, merges at St. Charles County, sometimes drowning the portion of it that is floodplain.

In a different sense, St. Charles County is being flooded with new development and traffic. It is on the northern cusp of metropolitan St. Louis, directly across the Missouri River from the city of St. Louis, the destination of the legions of commuters that clog the river crossings twice every working day. St. Charles is the fastest growing county in Missouri with a 1992 population of more than 226,000 people, an increase of over 50 percent since 1980. Much of the county's recent development has taken the form of new subdivisions, shopping centers, and office complexes in the Interstate 70 corridor west of St. Charles City. Most of this western portion of the county is upland, dissected by creek ravines and largely out of the direct reach of the 1993 floods, although traffic tie-ups were a major problem during the flood.

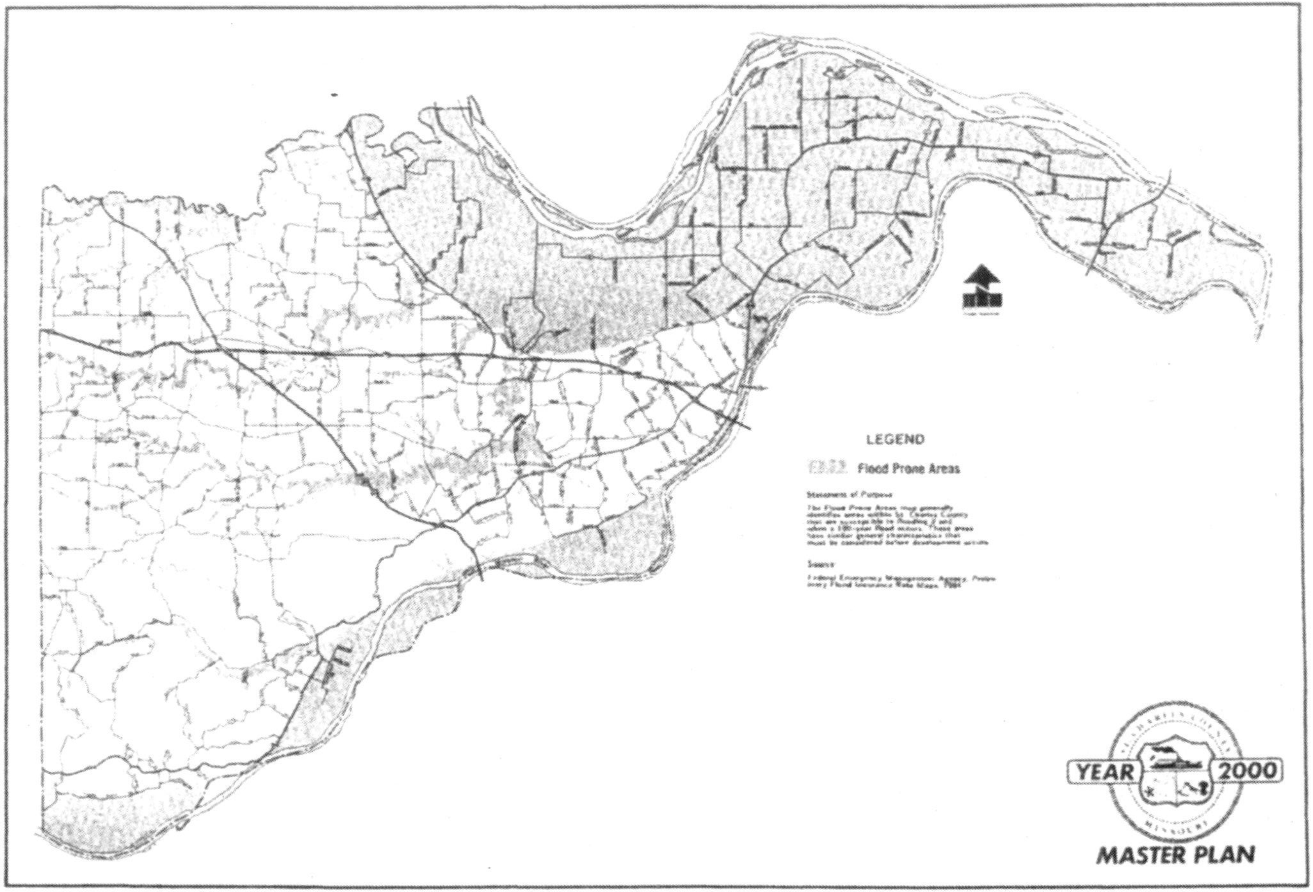

FIGURE 7-2 Floodplains of St. Charles County, Missouri, showing the approximate area flooded in 1993. (Source: St. Charles County)

St. Charles County contains several interior drainage basins, including the Femme Osage, Dardenne and Peruque Basins, and the Cuivre River Basin. The last three drain northward into the Mississippi River while Femme Osage Creek flows southward into the Missouri River. The floodplains along the interior creeks are vulnerable primarily to flash flooding brought on by local weather events, while the two mainstem river floodplains are threatened by gradual riverine overflow caused by large-scale midcontinental weather conditions.

The flood risk in St. Charles County is intensified by bluffs and levees lining the mainstem rivers. High bluffs form the northern bank of the Mississippi in Illinois, forcing the river southward into the broad flat floodplains of St. Charles County at times of high water. The Missouri west of St. Charles City is contained between parallel lines of bluffs that confine flooding to "bottoms" that lie between them on either side of the river. The bottoms north of Missouri lie in St. Charles County while those south of the river lie in St. Louis County. Portions of the bottoms are leveed to permit industrial development, which further constricts the natural floodplain. A levee designed to withstand a 100-year flood at Chesterfield, Missouri, just upstream and across the river from St. Charles County, failed in the 1993 flood, leaving an airport and industrial zone under water for several months.[8] Downstream from the confluence, large levees along both sides of the Mississippi at St. Louis and East St. Louis, Illinois, create a bottleneck and force floodwaters to back up into St. Charles County. These natural and manmade features, combined with the size of the two rivers, make St. Charles County one of the most frequently and severely floodprone jurisdictions in the country.

The Mississippi and Missouri Rivers, although often inflicting damage on the county, have also been very beneficial. St. Charles City, the county seat, is an old river port whose livelihood historically was dependent on the navigability of the river system. Also, the frequent flooding has created large areas of very fertile farmland in the St. Charles County floodplains through the deposition of sediments eroded from areas farther upstream. Also, the rivers and their floodplains have great recreational potential. Each year, tens of thousands of people enjoy boating, fishing, hunting, hiking, and bicycling on and along the rivers. The Katy Trail State Park provides a greenway and bike trail along the north bank of the Missouri River following a former railroad right-of-way across St. Charles County.

The high land values in the upland portions of St. Charles County contrast dramatically with the more sparsely settled bottomlands. The Missouri River bottoms contain only a few farm houses and associated agricultural structures. The Mississippi floodplain is predominantly farmland scattered with small subdivisions of modest-cost homes, fishing camps, and mobile home parks. The municipality of Portage Des Sioux and the unincorporated community of West Alton are both in the floodplain. In the 1950s and 1960s, a total of 28 subdivisions were platted along the Mississippi River. A large percentage of the lots in these subdivisions were developed as seasonal recreational housing. Eventually, many of these changed from seasonal occupation to year-round homes. By 1993, an estimated 4500 structures were located in the Mississippi floodplain within the county.

St. Charles County has a long history of flooding. Since joining the National Flood Insurance Program (NFIP) in 1978, the county has experienced four floods of a magnitude of 5 percent annual chance (20-year flood) or greater, in 1982, 1986, 1993, and 1995. For the period 1978 to 1994, the county ranked fifth among all jurisdictions in the nation in terms of repetitive flood insurance losses, with 4524 claims filed on 1345 St. Charles County properties totaling $54 million in NFIP payments.[9] (This of course does not reflect the hundreds of other structures not insured under the NFIP that also have been flooded repeatedly.)

Flooding is more frequent on the Mississippi than on the Missouri. Some of the subdivisions on the Mississippi River have been flooded once every two to three years. However, when flooding does occur on the Missouri, it is more devastating to the farmers due to the large amounts of sand carried and deposited by the flood waters. After the 1993 flood, some farmland was covered by six feet of sand making the acreage worthless for farming.

Flash flooding is frequent on the interior creeks. These creeks are surrounded by subdivisions, increasing the damage potential of any flood. The subdivisions also exacerbate flooding by increasing the amount of runoff and by installing inadequate stormwater detention basins. While flooding in these areas creates many problems for the county and a number of flood insurance claims, the percentage of federal costs is minimal in comparison with the amounts spent in the mainstem river floodplains.

Agriculture dominates the mainstem floodplains with a majority of the land included in working family farms. Some are protected by private levees. The remainder are either unprotected or within one of several levee districts. The largest of these districts, the Consolidated North County Levee District (CNCLD), is located in the eastern section of the county in the area threatened by both the Mississippi and Missouri Rivers. The construction of levees is controlled by the U.S. Army Corps of Engineers (USACE) and, in the case of levees rebuilt with federal disaster assistance, by the Federal Emergency Management Agency (FEMA). Some levees are built privately to create waterfowl habitat for duck hunting clubs. These structures tend to exacerbate minor floods by decreasing the amount of floodplain available and shunting water onto neighboring land and roadways.

Various commercial uses have developed in the floodplain including farm-related businesses, pipelines, power plants, marinas, and other recreational enterprises. Some of these may be hazardous in the floodplain due to the high level of pollutants associated with them. The pipelines carry a variety of petroleum products.

THE 1993 FLOOD

The flooding of 1993 in St. Charles County was unprecedented, as it was for much of the Upper Midwest, in terms of both record-setting water depths and the duration of the flooding. The river gauge at St. Louis, just downstream from St. Charles County, recorded the highest-ever peak stage (49.47 feet) although the peak discharge of about one million cubic feet per second was not an all-time record.[10] St. Charles County was twice declared a disaster area by President Clinton, first on May 11 and again on July 9. Portions of the county along the Mississippi River were under water from early spring until late fall. Some homes that were flooded in April could be reached only by boat for the next six months (Figure 7-3).

FIGURE 7-3 A breached agricultural levee along the Missouri River in St. Charles County. (Photo by R.H. Platt)

The 1993 floods had their inception in 1992. The fall of 1992 was rainy, saturating the ground and keeping river levels high. Heavy winter snows over all the Midwest increased the amount of water available within the Upper Mississippi River Basin. Flooding in the spring along the Mississippi River is a common occurrence in St. Charles County. However, in 1993 the floodwaters were higher than usual. All of the riverfront subdivisions along the Mississippi were flooded. Residents farther inland were also forced to evacuate. The amount of damage prompted the May 11 presidential disaster declaration for St. Charles and seven other Missouri counties.

During May and June, the Mississippi River receded slightly allowing some residents to begin repairing their homes. But in late June, a stalled Bermuda High over the southeastern United States and an abnormally cool air mass over the northwestern and north central United States resulted in what seemed to be almost continuous rainfall over the Missouri River Basin and the upper Mississippi River Basin.

Rainfall amounts in July across the upper Mississippi River Basin and the Missouri Basin exceeded four times normal amounts, breaking all records. A second presidential disaster declaration was issued on July 9, 1993, for 104 counties in Missouri. Ultimately, a total of 525 counties in nine midwestern states were declared disaster areas due to the 1993 flood, the most widespread natural disaster in U.S. history. The level of flood damage continued to rise after the declaration. The flood crest did not reach St. Charles County until August 1, 1993, nearly three weeks after the July 9 declaration.

At the crest of the flood in St. Charles County, nearly 30 percent of its land area or about 128,000 acres was under water. Flood waters from the Mississippi and the Missouri Rivers had covered the 100-year floodplains and caused back-flooding along the Dardenne, Femme Osage and Peruque Creeks, and the Cuivre River. The confluence of the Mississippi and Missouri Rivers temporarily moved 24 miles upstream into St. Charles County, overtopping levees and inundating the communities of West Alton and Portage Des Sioux.

Farmers in St. Charles were hit several ways by the 1993 floods. Crops planted by farmers in the flooded areas were ruined and the length of the flood precluded any replanting for the 1993 growing season. Other farmers who had a crop were also hurt by the flood when barge traffic was halted and bridges closed. A crop in the field is worthless if it cannot be delivered. Also, many farmers lost their homes and farm buildings to the flood and were displaced for months in great uncertainty about both their homes and their farm operations.

The high water overtopped all the agricultural levees in St. Charles County. The levees were damaged by scour, by wave action, and in many places by actual removal by the rushing water. More than 10 deep scours or "blue holes" were created, forcing the levee districts to build ring levees. The cost of repair was more than what many of the small levee districts and individual farmers had ever expected. As the water rushed over and through the levees, it scoured the farm-land behind them, removing topsoil and structures. In some places, gouges in the earth up to 60 feet in depth were created. The water also left huge amounts of debris behind, primarily sand and trees and remnants of houses, mobile homes, and vehicles.

At the Jacobs farm in southwestern St. Charles County along the Missouri River, the Augusta Bottoms Levee was overtopped near the farmstead. Due to the rising water behind the levee, the Jacobs family had removed all that they could from their home and barns. At the worst, the family had expected to come back to a severely flooded home. In the end, they came back to nothing. The floodwaters had completely demolished a two-story residence and outbuild-ings. The Jacobs's had only recently purchased the farm. With an existing loan on the property, the huge amount of damage and the combination of no crops and sand deposits, rebuilding was not an option, even with assistance from the Commodity Credit Corporation of the U.S. Department of Agriculture. Fortu-nately, the land was purchased by a neighboring farmer and the one-acre farm-stead was purchased as part of the St. Charles County buyout program (described below).

Overall, the flood damaged more than 4300 structures in the county and forced approximately 2000 families from their homes and farms, including some 400 families in three mobile home parks.[11] Total losses in St. Charles County alone were estimated to exceed $160 million, including agricultural as well as residential and commercial damage. The extent and duration of the flood and the resulting level of damage surpassed any previous experience of county officials.

Damage Assessment: Intergovernmental Confusion

St. Charles County responded to the disaster by developing a three-step process to guide damage assessment and the rebuilding process. It consisted of (1) dam-age inspection and assessment, (2) homeowner review, and (3) negotiation to expedite the repair or removal of flood-damaged structures in the floodplain. As soon as the floodwaters began to recede, damage assessment teams from the St. Charles County Building Department began to inspect flood-damaged homes to determine the level of damage. The total cost of the damage was compared to the market value of the structure as listed in the county tax records. If the level of damage was 50 percent or more of preflood market value, the building was tagged with a bright orange "condemned" placard. If the damage level was more than 25 percent but less than 50 percent, the building was tagged with a white "uninhabitable" placard. Structures that suffered less than 25 percent damage were tagged with a green "safe" placard.

All the placards included a telephone number to call for rebuilding options and other information. From this number, homeowners were directed to call the St. Charles County Planning Department to schedule an appointment with a planner. The planner would then review the damage assessment with the homeowner and explain what options were available.

The first weeks of inspections and reviews were a time of adjustment for all concerned. For the homeowners, it was an adjustment to the fact that the county was planning to enforce the regulations of the National Flood Insurance Program (NFIP). Some inspections had been done in June when the flood on the Mississippi River had receded slightly. However, these were limited in nature and not very detailed. Many residents quickly realized rebuilding would be at least very difficult and at worst impossible.

For the Building and Planning Departments of St. Charles County, it meant frequently adjusting the procedures and forms used. However, it also meant adjusting personal lives and perceptions of safety. The damage assessment process became an ongoing task of developing and refining the inspection and review procedures. The concept of what exactly constituted damage and how to resolve differing levels of damage changed almost weekly. This created extreme levels of stress for everyone involved—homeowners and public employees alike.

Many homeowners became very hostile to the damage inspectors, office staff, and the planners. The realization that the cost of rebuilding their home might be financially impossible caused many to verbally abuse anyone associated with the damage assessment process. On several occasions, sheriff's deputies were sent with the damage inspectors to a residence or called into the Building and Planning Departments to confront or remove an agitated citizen. Several inspectors and members of the staff received death threats.

Due to the level of hostility, no inspector ever went alone to a property and all vehicles had radios. The damage assessment office was in the St. Charles County Emergency Management Office (a bomb shelter), which was located inside the jail accessible only by a keypad security door. The planning and building departments were in temporary offices in the St. Charles County Court Administration Building, which had security guards and metal detectors. While these precautions made the employees feel a certain level of security, the fear of verbal and physical assaults never disappeared.

The level of tension was exacerbated by the multiplicity of responders to the disaster. Adjusters working for the National Flood Insurance Program gained rapid access to the damaged homes. Charitable organizations were conducting their own damage inspections at the same time. Then the county damage assessors arrived on the scene. All this created an incredible amount of confusion. Residents were often unsure which damage assessment was accurate. The insurance adjuster's estimate frequently exceeded the market value of the structure and was therefore to the owner's benefit. The charity's inspection was to assess assistance needs. The county's assessment often led to the condemnation of the structure. The county was supposed to be the final authority on damage levels and reconstruction options. However, homeowners often stated they had already been inspected and been told to start rebuilding, not realizing that the county had yet to visit their property.

The NFIP requires that structures that are "substantially damaged" may be rebuilt, if at all, only in compliance with currently applicable elevation and other floodplain management regulations. "Substantial damage" is defined as "damage of any origin sustained by a structure whereby the cost of restoring the structure to its before-damaged condition would equal or exceed 50 percent of the market value of the structure before the damage occurred."[12] This necessitates the determination of preflood value of each structure to determine those substantially damaged, which proved no straightforward task.

If a damage level of 50 percent or more meant that a residence was condemned and therefore must be elevated or demolished, then *50 percent of what* became the crucial question. The county used the fair market value of the structure as determined by the county Tax Assessor's Office. Insurance adjusters often used the structure's replacement value or the cost of completely replacing the structure using current material and labor prices to determine the level of damage. This meant the county might use a market value of $24,000 while an insurance adjuster used a replacement value of $50,000. Obviously, the latter approach would result in fewer structures being declared "substantially damaged."

This conflict became even more severe for the county when the Federal Emergency Management Agency (FEMA) stated in October 1993 that communities could use *either* the market value or the replacement value. St. Charles County was then placed in an extremely difficult position. Nearly a third of the damage inspections and the assessment reviews had been completed. Many people had been told that the only options available were demolition or elevation. For the elderly and those of low-income these were devastating responses. The decision by FEMA to allow the use of replacement value forced the county to review hundreds of completed files and to revise a number of them.

FEMA's decision also forced the county to once again redo the damage assessment process. A totally new method of determining repair costs when using replacement values had to be created. The county eventually used construction costs from the Marshall-Swift Estimating Manual when determining the level of damage for replacement value as well as the existing method developed for the market value. In its final form, the damage assessment process required that a structure be damaged by less than 50 percent by *both* market value and replacement cost before a permit to repair without elevating would be issued.

Another problem was the speed with which both the federal government and charitable organizations worked to assist the flooded homeowners to repair their homes. Far too often, homeowners were provided with funds needed to start repairs before the county had inspected the structure or issued a permit. In the rush to make flood victims whole again, the role of the local government was often overlooked. This rush to provide assistance forced St. Charles County to take the radical step of shutting off all electricity to certain flooded areas, thereby increasing the bitterness of affected flood victims. Only by taking this step, however, could the county damage assessment be completed before homeowners began repairs.

A final problem was the inability of local government officials to obtain precise information on the flood insurance claims paid to the owners of damaged structures, which sometimes exceeded the estimated level of damage. In other words, property owners would claim the maximum possible level of damage from the NFIP but then turn around and argue that their actual damage was less than 50 percent of fair market value or replacement cost. Federal regulations allow homeowners to dispute the level of damage determined by the local official. If the homeowner submitted a bid proving the level of damage was less than 50 percent, then the structure did not need to be elevated. The bids for repair were often much less than the insurance claim. In some cases, as much as 75 percent less. Also because homeowners could charge only $4.35 per hour for their own labor, they could propose to do the work themselves at very low cost, thus reducing the damage estimates. Without access to accurate information, the county often seemed to be assisting some homeowners in a fraudulent scheme to avoid a substantial damage determination. Access to the amount of the insurance claim would have bolstered the county's claim that these structures were damaged by more than 50 percent and therefore required elevation.

The combination of all these problems made the task of determining the level of damage and controlling repairs very difficult for St. Charles County and hundreds of other jurisdictions throughout the Midwest. It also increased the level of confusion and stress experienced by the flood victims. Many homeowners, particularly the elderly, were totally unable to decide what should be done next when faced with the bewildering array of government agencies and assistance programs that existed to assist flood victims.

The Buyout Program

In the fall of 1993, the primary source of federal funding to assist in the acquisition of flood-damaged properties was Section 1362 of the National Flood Insurance Program, which was limited in both applicability and authorized level of funding. The county submitted an application to FEMA in September to initiate some limited acquisition under the NFIP program. This proposal was abandoned when a larger buyout program was announced in October that would use hazard mitigation grant funds under Section 404 of the federal Disaster Assistance Act (Stafford Act).

Based on early assessments of the nature of damage caused by the 1993 Midwest Flood, the Clinton administration asked Congress to augment available Section 404 funding to help as many victims as possible to move to safer loca-

tions. In December 1993, Congress adopted the Hazard Mitigation and Relocation Assistance Act (PL 103-181) for this purpose. St. Charles County was a
prime candidate for the implementation of this approach since it had already
been identified as having one of the highest rates of repetitive flood insurance
losses in the country. Also, most of the dwelling units in its floodplains were of
modest value, consistent with the finding of the Interagency Floodplain Management Review Committee that housing damaged in the 1993 floods was generally valued at $25,000 or less.[13] The state of Missouri ultimately received close
to $100 million, which flowed through the state to local communities. This
included $30 million from the FEMA hazard mitigation grant program (Section
404 of the Stafford Act), $28 million from FEMA Section 406 disaster assistance
funds, and $42 million in community development block grant (CDBG) funds
from the Department of Housing and Urban Development (HUD). Overall, the
Missouri buyout program was the nation's largest with 44 community projects
involving the purchase of approximately 5300 properties.[14] The largest local
buyout program in the nine-state region affected by the flood was conducted by
St. Charles County, which eventually purchased about 1620 properties—almost
one-third of the total for Missouri—including 548 single-family homes, 209
mobile homes with pads, 598 mobile home pads without homes (some
removed), 95 vacant lots, and 10 commercial properties.[15]

The buyout program was not universally welcomed. Many residents in the
flooded areas viewed the buyout program as a way to force homeowners out of
the floodplain and to leave it for the "tree huggers." Informational meetings held
to explain what the county was proposing often became confrontational. County
staff attempting to facilitate such meetings often feared that the attendees were
merely one step away from physical violence.

The buyout program in St. Charles County was directly administered by
RJK, Inc., a private contractor paid out of federal funds. The contractor was
responsible for all negotiations with potential sellers, for obtaining necessary
appraisals, for preparing legal documents, and for clearing the property of structures after title was conveyed to the county. Neither the county nor the contractor was responsible to assist sellers in finding a new home. PL 103-181 specifically exempted the buyout program from the federal Uniform Relocation Act of
1970. However, the contractor was situated in the same office space as state and
county social service agencies, providing one-stop shopping for various sources
of assistance.

The buyout program was complex. As stated earlier, federal funds were
derived from both FEMA and HUD. Rules of eligibility differed widely between
the two agencies, and each property acquired had to satisfy one or the other.
Technically, the federal share of the buyout was 75 percent, with the balance to
be provided by the state or local jurisdiction, or both. However, in St. Charles
County as elsewhere in the buyout program, the "nonfederal" share was actually

paid with other CDBG funds. (The latter were considered nonfederal because the local recipient had discretion as to how they should be used.) So the entire cost of the buyout program in St. Charles County was essentially federal.

Soon after the flood, various federal agencies paid nearly $24 million in flood insurance payments, disaster assistance, and SBA disaster loans to owners of properties in St. Charles County that were later acquired (Table 7-2). (This was a small proportion of total federal disbursements in the county.) Where such funds had not yet been used for the repair of the flooded properties, they were deducted from the appraised preflood value of the property to avoid duplication of benefits. This provision reduced the total amount paid to sellers by about half, from $20.5 million to $10.2 million. Administrative costs amounted to $3.5 million, yielding a total of about $13.6 million paid by FEMA and HUD to buy out 1374 properties in St. Charles County, including vacant lots and mobile home pads. This averaged $9,800 per property. The cost of the 640 dwelling units acquired (fixed and mobile) averaged about $21,000 each (Table 7-3).[16]

Acquired properties were restricted to open space uses under FEMA rules. But the parcels tended to be very small and widely scattered across the floodplain, except where contiguous blocks of mobile home lots were acquired. The question thus arose as to how should the acquired parcels be managed and by whom? Once lots were cleared of structures, the county offered to lease the individual parcels for $1 per year to adjoining property owners. The leases consisted of a one-year maintenance agreement that required the lessee to use the land only as open space or for farming. Of course, all acquired lots were withdrawn from the property tax roll, whether or not leased to abutting owners.

Sellers had to agree not to use the funds to acquire new property in a floodplain, but otherwise were on their own. In light of the vast difference in real estate prices between the bottomlands and the upland portions of the county, one may conjecture that many of the sellers left St. Charles County. Some are likely

TABLE 7-2 Federal Flood Costs on Properties in St. Charles County Purchased After the 1993 Flood

NFIP structural damage claims	$10,312,733
NFIP contents claims	$2,987,797
Emergency repairs	$836,391
Individual and family grants	$419,797
Mobile home NFIP claims/disaster aid	$5,169,872
Small Business Administration loans	$3,804,390
NFIP loss processing costs (est.)	$399,000
Total 1993 Federal Costs on Buyout Properties	$23,929,980

Adapted from Missouri State Emergency Management Agency, "Out of Harm's Way: The Missouri Buyout Program" Jefferson City, Mo.: MSEMA, 1995

TABLE 7-3 St. Charles County Buyout Program

Total Fair Market Value of Purchased Properties	$20.5 million
Duplication of Benefits (NFIP, SBA, Disaster Aid)	(−$10.5 million)
Actual Purchase Price (1347 parcels including vacant mobile home pads)	$10 million
Administrative Costs	$3.6 million
Total Cost of St. Charles Buyout Program	$13.6 million
Cost per Purchased Property ($13.6 million/1374 parcels)	$9,800
Cost per Dwelling Unit Purchased ($13.6 million/640 dwelling units)	$21,250

Adapted from Missouri State Emergency Management Agency, "Out of Harm's Way: The Missouri Buyout Program" Jefferson City, Mo.: MSEMA, 1995

to be still living in the bottomlands to which many long-time residents hold a strong attachment. Neither the federal government nor the county have attempted to monitor where relocatees went or how they used buyout payments and other federal benefits.

Replacement Housing

Moving to higher ground in St. Charles County inevitably involves higher housing costs. In an attempt to assist the many families who were displaced by the flood and the buyout program, St. Charles County, in conjunction with developers, proposed several projects to create lower-cost replacement housing. The county submitted an application to the state of Missouri for $3.85 million in late 1993 to assist in the relocation of flood victims. Primary considerations for the relocation housing request were to retain the families in the Orchard Farm School District and maintain access to local shopping and other services. While scattered sites had been considered, a single subdivision was determined to be the most cost-effective method of developing replacement housing.

Approximately $2.3 million of the requested funds would have been used to acquire and expand an existing mobile home park called Deerfield. The park was located within the corporate jurisdiction of the city of St. Charles, which exercised control over zoning and utilities pertaining to the site. This portion of the proposal was rejected by Governor Carnahan because Deerfield had been partially flooded and the construction of a levee would be required.

The remaining $1.5 million was approved. These funds were to be used to subsidize the public work costs of a 100-home subdivision. The funds were to be provided by the CDBG Program and the Missouri Housing Development Commission (MHDC). Several sites were considered for possible development.

The first site was proposed in early 1994 near the area of three mobile home parks targeted for acquisition along Boschertown Road. The area was located in unincorporated St. Charles County but bordered several subdivisions located in the city of St. Charles. The site was abandoned after loud protests from nearby subdivision residents resulted in the refusal of the city to extend sewer and water utilities to the area. More than 450 residents had signed petitions in opposition to the development of "subsidized" homes. A second site was investigated in the summer of 1994. This site was also located along Boschertown Road and was also abandoned for reasons similar to the first—namely, loud community protest and unwillingness of the city to extend utilities.

A third site in an area called Orchard Farm was considered in early 1995. Orchard Farm consists of fewer than 50 homes along Highway 94N, approximately 10 miles north of the city of St. Charles. The 48-acre parcel to be developed was located in the 500-year floodplain. A maximum of 110 homes was proposed with utilities to be provided by the city. By July 1995, protests from the surrounding homeowners and other interested individuals resulted in the withdrawal of the CDBG funds from the project.

By the end of the buyout program, no replacement housing was built in St. Charles County. Families displaced by the 1993 floods were forced to secure permanent housing through their own efforts or with the assistance of the local social service agencies. The primary reason for the failure to develop replacement housing was the unwillingness of current residents to allow what was perceived as "lower class" subsidized housing to be built in or near their neighborhoods. The county, state, and federal governments proved unable or unwilling to withstand such opposition or to use eminent domain powers to develop alternative sites. This proved to be a major failing of the buyout program in St. Charles County.[17]

In May 1995, floodwaters returned to St. Charles County inundating many of the same areas as in 1993 but causing a lot less damage. Federal and state disaster officials proudly compared statistics for the two floods (Table 7-4).

After spending some $15 million in St. Charles County to buy up 1437 parcels and 640 dwelling units (including mobile homes), it is understandable that the data in Table 7-4 should be gratifying to advocates of the buyout program. However, the two flood events were not physically comparable. The flood

TABLE 7-4 Comparative Disaster Costs in St. Charles County: Floods of 1993 and 1995

	1993	1995
No. of Applicants for Federal Disaster Assistance	4227	333
Disaster Housing Costs	$8.3 million	$204,000
Individual/Family Grants	$5.8 million	$11,000
SBA Disaster Loans	$11.8 million	$67,000

Adapted from Missouri State Emergency Management Agency, 1995, "Out of Harm's Way: The Missouri Buyout Program." Jefferson City, Mo.: MSEMA.

crest in the 1995 event was seven feet lower at St. Charles County than in 1993[18] and the flood duration was a matter of days rather than months. The buyout program in the county at most would have eliminated 600 household units applying for federal assistance, leaving some 3200 claimants from 1993 still in the area who did not file a claim in 1995. It was, simply, a much smaller and briefer flood. One wonders whether a federal disaster declaration was really needed for the 1995 event.

To be sure, the owners of the more than 640 dwelling units that were purchased and demolished or removed from the floodplain were presumably in a safer location ("out of harm's way") than in 1993. And the 1437 acquired lots and mobile home pads were no longer eligible for future residential use, thus reducing potential future vulnerability. An unknown number of homes were elevated in the rebuilding process as a result of a substantial damage determination; this could have accounted for a portion of the nonclaimants in 1995. St. Charles County was indeed somewhat less at risk from a repetition of the 1993 flood. But how much less, at what cost, and to whom?

The Larger Context: When Is Enough, Enough?

No one actually knows the total costs to the federal government of the 1993 flood in St. Charles County, or any other particular jurisdiction for that matter. Inquiries for this study disclosed that there is no central database that records the expenditures by all federal responding agencies for specific disasters and governmental units. FEMA tabulates its own expenditures under the NFIP and the Stafford Act by state declaration, but does not record outlays of other agencies such as HUD, SBA, USDA, and the Army Corps of Engineers. Nor do crude tallies of program outlays, if available, reflect the actual cost to the federal government. Insurance costs are defrayed to some extent by premium revenue. Disaster loans are subject to repayment, with the federal subsidy measured by the present value of the interest rate subsidy over the life of the loan. (In the event of default on repayment of disaster loans, SBA may foreclose on the borrower's property, although the government's claim is usually subordinate to those of mortgage lenders, if any.)[19]

As discussed in Chapter 1, overall federal disaster assistance, including grants, insurance payments, and loans, has expanded dramatically in the 1990s over prior experience. One indicator of the general size of federal disaster outlays is the upward trend in congressional supplemental disaster appropriations. These ad hoc appropriations augment FEMA's regular budget of about $300 million per year and also fund response efforts by many other federal agencies and programs. Supplemental disaster appropriations between 1980 and 1988 totaled about $7.2 billion, while between 1988 and 1994 they amounted to $17 billion (both figures in constant 1993 dollars).[20] In June 1997, Congress made another supplemental appropriation of $5.1 billion for flood relief in 23 states, including particularly those struck by the Red River Flood in the Dakotas and Minnesota that year.

P.L. 103-75 signed by President Clinton on August 12, 1993, at the height of the Midwest Flood, appropriated $6.3 billion for anticipated federal costs of that disaster. Of that amount, $2 billion was allocated to FEMA "for the Midwest floods and other disasters" and the rest to other agencies, mostly for flood-related activities. The Galloway Report estimated that total federal costs relating to the Midwest Flood would amount to about $4.2 billion in direct federal expenditures (Table 7-5), $1.3 billion in payments from federal insurance programs, and more than $621 million in federal loans to individuals, businesses, and communities.[21]

Reflecting FEMA's emphasis on mitigation under the Clinton administration, Congress on December 3, 1993, amended the Stafford Act to raise the federal share of hazard mitigation measures from 50 percent to 75 percent, and to raise the percentage of federal disaster assistance available for mitigation projects from 10 percent to 15 percent.[22] Under this and other congressional authorities, about 12,000 floodprone properties were acquired at a federal cost of about $375 million (including about $152 million in FEMA hazard mitigation grant funds).[23] This was an unprecedented level of effort to remove structures and floodplain residents from floodplains in St. Charles and elsewhere in the nine-state region. Yet it comprised only 3.6 percent of total direct federal disaster costs of $4.2 billion stemming from that flood. The buyout program was in fact eclipsed by the vastly greater federal funding devoted to replacement of infrastructure (e.g.,

TABLE 7-5 Direct Federal Expenditures for the Midwest Flood of 1993 in Millions of Dollars (excluding federal insurance and disaster loan programs)

Department of Agriculture (USDA)	$1,699.9
Federal Emergency Management Agency (FEMA)	$1,098.0
Department of Housing and Urban Development (HUD)	$500.0
Department of Commerce	$201.3
Army Corps of Engineers	$253.1
Department of Education	$100.0
Department of Labor	$64.6
Department of Transportation	$146.7
Environmental Protection Agency	$34.0
Department of the Interior	$41.2
TOTAL FEDERAL EXPENDITURES	$4,138.8

Source: Interagency Floodplain Management Review Committee, 1994. Sharing the Challenge: Floodplain Management into the 21st Century. Washington, D.C.: USGPO, Table 1.4.

highways, levees, sewer and water systems) and to myriad other federal disaster relief programs.

Despite widespread approval of the buyout approach, a number of post-flood critiques of federal policies concerning the Midwest Flood have questioned the wisdom of the vastly expanded scale of federal disaster assistance reflected in that disaster. The Galloway Report, the authoritative post-flood review of federal policy commissioned by the White House, warns against "moral hazard" whereby individual incentives to guard against disaster are lessened by overzealous government assistance programs. In other words, as discussed in Chapter 1, the sheer enormity of federal largesse following disasters may encourage local governmental and private decisions that increase rather than reduce vulnerability to floods.[24]

The same message is conveyed by the House Bipartisan Task Force on Disasters report of December 14, 1994.

> If homeowners mistakenly believe that the Federal Government will rebuild their homes after a natural disaster, they have less incentive to buy all-hazard insurance for their homes. If state and local governments believe that the Federal Government will meet their needs in every disaster, they have less incentive to spend scarce state and local resources on disaster preparedness, mitigation, response, and recovery. This not only raises the costs of disasters to federal taxpayers, but also to our society as a whole, as people are encouraged to take risks they think they will not have to pay for.[25]

James Wright, former head of the Tennessee Valley Authority Floodplain Management Office, in reflecting on the 1993 Midwest Flood, calls for greater assumption of individual responsibility:

> Various legislation (e.g., flood insurance) calls for individuals to bear the full cost for their decision to utilize flood-hazard locations. This . . . is not happening. Strict enforcement of present measures is needed, with strong penalties for noncompliance. Taxpayers should not be expected to bail out uninsured flood victims. Those who choose to continue to live in hazardous areas with full knowledge of the risk should be expected to suffer the consequences. . . . The federal government must integrate its disaster assistance policies and programs with those that promote longer-term solutions to flood problems. Relief should be linked to responsibility. . . . [M]any citizens simply expect public aid when calamity strikes. With 50 federal programs to supplement

their courage, people have every incentive to build in floodprone areas and leave the risk to Uncle Sam.[26]

In September 1994, the Lincoln Institute of Land Policy in Cambridge, Massachusetts, convened a symposium entitled "Community Land Policy and River Flooding: The Great Flood of 1993." The participants included invited public officials from 33 municipalities affected by the flood, as well as professionals from a variety of disciplines. According to the meeting summary, federal disaster assistance was viewed by some participants as an "addiction":

> Many participants at the Lincoln Institute conference suggested that the federal government should restrict post-disaster support for communities that fail to plan ahead for floods. Some participants extended the principle to local governments and nongovernmental organizations as well, calling for all sources of disaster assistance to practice "tough love." They compared federal, state, and even private disaster assistance to an addiction from which floodplain communities and residents would not recover if it was always easier to ask for help after a flood than to get out of harm's way ahead of time.[27]

Finally, Howard C. Kunreuther and Gilbert F. White, in reflecting on the Midwest Flood recovery, observe:

> There has been a distinct shift in public attitude toward disasters in recent years. Following the San Angelo tornado in 1953, federal representatives advised local school officials to call for federal funds to repair a school damaged by the twister. They were reluctant to even make a request for the repair cost that was not covered by insurance. Eventually they did and it was promptly granted. One year later they returned the money in an act of conscience. It is hard for us to imagine such behavior today. In fact, there is now an expectation from taxpayers and hard-hit communities that the federal government will come to the rescue with grants and low-interest loans.[28]

CONCLUSION

The foregoing case study discloses a schizophrenic character to the recovery of St. Charles County and the Midwest generally from the 1993 floods. On the one hand, mitigation in the form of the buyout program played a greater role than in any previous flood disaster. Congress approved new levels of funding and a higher proportion of federal participation for the purchase of chronically flood-threatened properties. The acquisition of some 1400 parcels of land in St. Charles County, along with the demolition or removal of about 600 dwelling units, was a tangible accomplishment of the Clinton administration's commitment to mitigation. The buyout approach received widespread approval from public officials, the press, and hazard specialists.

On the other hand, the critiques just summarized by some of the nation's leading experts on flood policy look beyond the $375 million federal buyout program to the $4.2 billion overall estimated cost of federal disaster assistance in the nine-state region. It is noted that in many cases, no state or local contribution was required since nonfederal share of buyouts and certain other activities was provided through federal community development block grant (CDBG) funds administered through the Department of Housing and Urban Development. (And, as discussed in Chapter 1, it is very difficult even for federal government analysts to track down the exact allocation of CDBG disaster funds.)

The issue is thus presented as to whether the federal government has wisely assumed nearly 100 percent of the costs of disaster response and recovery, while concomitantly expanding the eligible types of assistance that it funds. Has disaster assistance in the words of the Congressional Research Service assumed the status of an "entitlement"?[29] If so, is this justified for disaster victims, regardless of need, while nothing of a similar magnitude and level of federal beneficence is offered to low-income residents of chronically hazardous neighborhoods of central cities? At a time when Congress seems intent on cutting federal contributions to health care, education, housing, and environmental programs, disaster assistance is unique in commanding almost automatic and unlimited bipartisan acclaim.[30]

Returning to St. Charles County, a number of issues deserve further consideration. First, while the public authorities involved can be justly proud of the buyout program, the results of this approach remain to be evaluated. For instance, what is known about the whereabouts of the individuals who sold their floodprone properties? Were the small amounts they received sufficient to help them find adequate, safer housing? What about those who did not sell? Why did they not take advantage of the buyout program? Did they elevate or otherwise floodproof their homes after the 1993 flood? How will the land acquired be utilized?

A second set of issues concerns the role of the county as the local land use planning and zoning authority. A critical and troublesome responsibility imposed on the county by the National Flood Insurance Program was the identification of "substantially damaged" structures that would have to be elevated if rebuilt. County officials were confronted with hostility and threats to their personal safety in making these determinations. But to make matters worse, FEMA changed the guidelines for "substantial damage" in the middle of the process, authorizing replacement cost in place of preflood value as the standard for determining whether a structure was damaged by more than 50 percent of such value. This undercut the efforts of the county to apply the more stringent standard of preflood value and threw many calculations into confusion and dispute. Evidently, FEMA and the White House were expressing solicitude for the victims and sought to ease the burden of mandatory elevation for as many homeowners as possible. In so doing, they made a difficult task even more onerous and presumably reduced the effectiveness of the substantial damage test as a mitigation measure.

The county's ill-fated effort to provide replacement housing for some of the households displaced from the floodplain highlights another weakness in the national policy on disaster recovery. While the federal government is expected to provide access to unlimited funding, it has virtually no influence over local land use and zoning practices. While rehousing flood victims would be consistent with the buyout approach, it proved impossible in St. Charles County to promote that worthy goal due to NIMBY ("not-in-my-backyard") objections from neighboring residents. FEMA made no effort to overcome this obstacle, choosing to remain aloof from local planning issues, even though its buyout approach created the need for alternative housing.

St. Charles County strongly discourages new development in its floodplains. This, however, was not an explicit requirement of the buyout program or federal assistance in general. A less proactive local jurisdiction could potentially accept the federal buyout funding and later allow sites not acquired to be developed, thus nullifying the effect of the buyout effort. Another practice commonly seen along ocean coasts is to permit (and perhaps encourage) the upgrading or replacement of existing structures in hazardous areas (albeit elevated) even as public money is used to remove other properties from developed status.

Alternatives to public purchase should have been more carefully evaluated. Public ownership of a patchwork of tiny lots and mobile home pads is not per se a desirable goal. The individual lots are removed from the tax base and from the oversight of the private owner and may attract litter and trash, as well as encroachment by neighbors. The results of St. Charles County's policy to lease acquired parcels to adjoining property owners for a dollar a year are not yet clear. If public acquisition is undertaken, it should be designed to ensure that holdout owners do not reap an undeserved windfall through de facto enlargement of their own properties at no cost.

The recovery from the Midwest Flood at St. Charles County reflected the best and the worst instincts of the American political system: compassion and avarice. The buyout program sought to help lower-income residents of the floodplain to relocate, an admirable purpose. Over one-third of the buyout budget, however, was paid to private contractors who administered the program on behalf of the county and state. Victims received sometimes pitifully small payments with which to relocate themselves and their possessions. Meanwhile, the cornucopia of federal disaster assistance paid for the restoration of public infrastructure, crop payments, social services, and loss of local taxes among a multitude of other benefits. Were these larger expenditures wisely employed? Should the immediate victims have received more, and the disaster recovery industry contractors much less? That question remains an enigma.

NOTES

1. Eberhard Czawa (1981). *Rivers of the World.* New York: Van Nostrand Reinhold, 2.

2. Anthony Trollope (1986). *North America.* New York: Da Capo Press, 146.

3. Rutherford H. Platt (1995). "The Mississippi River Basin: Crucible of National Flood Policies," in J. Gardiner, O. Starosolszky, and V. Yevjevich, eds., *Defence from Floods and Floodplain Management.* Dordrecht, The Netherlands: Kluwer Academic Publishers, 381–394.

4. Stanley A. Changnon, ed. (1996). *The Great Flood of 1993: Causes, Impacts, and Responses.* Boulder, Colo.: Westview Press, 4.

5. Interagency Floodplain Management Review Committee (IFMRC) (1994). *Sharing the Challenge: Floodplain Management Into the 21st Century.* Washington, D.C.: U.S. Government Printing Office, Table 1.1.

6. Geoffrey J.D. Hewings and Ramamohan Mahidhara (1996). "Economic Impacts: Lost Income, Ripple Effects, and Recovery," in Changnon, note 4, 205–217; Scott Faber (1996). *On Borrowed Land: Public Policies for Floodplains.* Cambridge, Mass.: Lincoln Institute of Land Policy.

7. IFMRC, note 5, 22.

8. Richard C. Shepard (1994). "Floodplain Development: Lessons Learned from the Great Flood of 1993," *Urban Land* 53(3): 19–24.

9. Federal Emergency Management Agency, unpublished data.

10. Mary Fran Myers and Gilbert F. White (1993). "The Challenge of the Mississippi Flood," *Environment* 35(10): 6–9, 25–36.

11. Miriam Anderson (1996). "Midwest Floods of 1993: St. Charles Experience," *Forum for Applied Research and Public Policy* (Fall issue): 129–132.

12. 44 *Code of Federal Regulations,* Sec. 59.1.

13. IFMRC, note 5, 7.

14. Missouri State Emergency Management Agency (1995). *Out of Harm's Way: Missouri's Flood Buyout Program.* Jefferson City, Mo.: MSEMA.

15. Data provided by RJK, Inc., for St. Charles County, as of September 1, 1995.

16. MSEMA, note 14. The total of 1374 properties used in this report and shown in Table 7-3 reflects an earlier enumeration than the total of 1620 properties referred to in the preceding note.

17. Elsewhere, as in Grafton and Valmeyer, Illinois, entire small towns were relocated out of the floodplain under community redevelopment projects funded by the federal government. See William R. Morrish, Carol J. Swenson, and Michelle M. Baltus (1994). *Recovery and Resettlement: A First Look at the Post-Flood Recovery Planning Issues in the Upper Mississippi River Valley.* Minneapolis: Design Center for American Urban Landscape, University of Minnesota.

18. Iver Peterson (1995). "In Flood Zone, Not as Bad as '93 Is Bad Enough," *The New York Times* (May 12), A1 and A11.

19. Becky Brantley, SBA Disaster Assistance Loan Officer, pers. comm. June 19, 1998.

20. U.S. Senate Bipartisan Task Force on Funding Disaster Relief (1995). *Federal Disaster Relief.* Washington, D.C.: U.S. Government Printing Office, Table 5.1.

21. IFMRC, note 2, 22.

22. P.L. 103-181, 42 USCA Sec. 5170c.

23. Stephen Braun, "Midwesterners Bailing Out of Floodprone River Plains." *Los Angeles Times* (July 14, 1995), A1 and A7.

24. IFMRC, note 5, 180.

25. U.S. House of Representatives Bipartisan Task Force on Disasters, *Report.* Washington, D.C.: mimeo, 1.

26. James M. Wright (1994). "Transcending the 1993 Midwest Flood: National Policy and Program Implications." Paper presented at the American Association for the Advancement of Science, San Francisco, February 22, 1994. See also James Wright (1996). "Effects of the Flood on National Policy: Some Achievements, Major Challenges Remain," in Changnon, note 4, 245–275.

27. Faber, note 6, 10.

28. Howard C. Kunreuther and Gilbert F. White (1994). "The Role of the National Flood Insurance Program in Reducing Losses and Promoting Wise Use of Floodplains." *Water Resources Update* (Spring): 31-35.

29. Keith Bea (1992). *FEMA and the Disaster Relief Fund.* Washington, D.C.: Congressional Research Service, 17.

30. In June 1997, a supplementary appropriation bill including over $5 billion for flood disaster assistance was vetoed by President Clinton due to unrelated riders attached by Republicans. After a frenzy of mutual scolding for being unfeeling toward the plight of the Red River flood victims, both sides quickly approved a clean bill a few days later.

CHAPTER 8

The Bay Area:
One Disaster After Another[1]

> A couple in their forties stood clutching each other on the fringe
> of the crowd. They were focused on a point in the inferno which
> was directly in our line of view. The flames licked and roared at
> some houses not yet on fire but oddly seemed to veer away from
> others. Trying to predict what would explode next was like try-
> ing to predict the path of a tornado. Suddenly the couple both
> yelled out: "Oh God, that's our house."
>
> Frances Rowe, "Burn, Baby, Burn," in Patricia Adler, ed.,
> *Fire in the Hills: A Collective Remembrance.* Berkeley:
> Patricia Adler (1992)

> We have several thousand properties in the hills that are over-
> grown. . . . We were surprised that the experience of 1991 did
> not spark more compliance. . . . It doesn't just affect the way they
> live—they are also posing a threat to their neighbors.
>
> Maura Smith, Oakland city planner,
> quoted in "'91 Fire Victims Slow to Learn."
> *San Francisco Chronicle,* October 29 (1993), A20

INTRODUCTION

San Francisco, the core city of the "Bay Area," is considered by many to be the
most beautiful and vibrant city in North America, renowned for its spectacular
views, its bracing air, its fogs, its Pacific Ocean cliffs, the spectacular San Francisco Bay, its striking architecture, and its cultural diversity. The larger Bay Area
reaches northward across the Golden Gate to opulent Marin County and the
wine country of the Sonoma and Napa Valleys. To the south lies the "Silicon Valley," seedbed of the global computer industry, while the East Bay urban corridor,
including Berkeley and Oakland, face San Francisco across the Oakland Bay
Bridge.

The Bay Area is designated by the Census as the "San Francisco–Oakland–San
Jose Consolidated Metropolitan Statistical Area."[2] It is the fourth largest urban

241

region in the United States, with 6.2 million people. The region extends approximately 90 miles from Santa Cruz in the south to Santa Rosa in the north, and about 30 miles from the Pacific Ocean to the city of Walnut Creek east of the Oakland–Berkeley Hills. The region contains nine counties and several hundred cities and towns that roughly encircle San Francisco Bay and its northerly extensions, San Pablo Bay and Suisan Bay

San Francisco and the Bay Area occupy a special status among American urban regions, both for their unique blend of spectacular natural and cultural amenities and for their chronic vulnerability to devastating natural hazards. Foremost among these is the risk of earthquake along any of several faults that interlace the region from north to south. In 1906, San Francisco was substantially destroyed by an 8.2 Richter magnitude earthquake followed by fires that engulfed the city from ruptured chimneys and exploding gas mains. On October 17, 1989, the 7.1 magnitude Loma Prieta Earthquake, centered 70 miles south of San Francisco, reminded Bay Area residents that another "Big One" may not be far in the future.

More localized physical risks include landslides, mudslides, flooding, soil liquefaction, and wildfire. During the 1990s, winter rains, compounded by the El Niño climatic anomaly, have triggered widespread flooding and mudslides. The spread of homes on the steep slopes and chaparral ecosystem of the East Bay hills has contributed to a series of destructive wildfires, most notably the Oakland–Berkeley firestorm of October 20, 1991. Compounding the effects of natural disasters is the potential failure of lifelines, including transportation, communications, water and sewer systems, dams, energy systems, food, and medical care delivery.

Like California in general, the Bay Area has learned much about coping with natural hazards through direct experience. Seismic building standards, first developed after the Long Beach Earthquake of 1933, have helped to promote safer construction practices in the region and state. The 1971 Alquist-Priolo Act required the preparation of seismic hazard maps, and in 1975 the program was expanded to require realtors to disclose such hazards to prospective buyers of certain types of real estate.[3] California has the nation's leading state emergency management office, and the Bay Area can draw on a variety of fiscal, technical, and intellectual resources to adapt its built environment and its population to the natural limitations of its site. But as elsewhere in the United States, the achievement of optimal mitigation approaches is tempered by the constraints of property rights and politics.

The Physical Setting

In contrast to the Atlantic shore, the U.S. Pacific coast offers few safe harbors, and thus few opportunities for the historic growth of great cities. Between Puget Sound to the north and San Diego Harbor near the Mexican border, there is only one major opening in the treacherous and magnificent succession of sea cliffs and coastal barriers that line the West Coast, namely the Golden Gate. This fabled strait, spanned 50 years ago by the Golden Gate Bridge, connects the open Pacific with the vast, sheltered San Francisco Bay. The Bay forms an immense estuary extending about 55 miles north to south. While much of the Bay's shoreline has been filled for urban development, its northern reaches and its extension into Suisan Bay retain important wetland habitat as nurtured by freshwater flow from the Sacramento River Delta. In the nineteenth century, the Sacramento River afforded access by riverboat to the interior Central Valley and the Sierra Nevada, thus promoting the growth of San Francisco as the gateway to a vast agricultural and mining hinterland. In the words of one of the city's early enthusiasts:

> San Francisco, . . . however ill-favored the site in some respects, seems topographically marked for greatness, rising on a series of hills, with a great harbor on one side, a great ocean on the other, mighty waters ever passing by to the outlet of the widespread river system of the country.[4]

The site of San Francisco outwardly appears to be highly favorable to urban settlement. It lies at the northern tip of a broad peninsula accessible by sea or by land from a region of fertile valleys and coastal hills to the south. The site of the city itself was originally bordered on the bay side by marshes that were easily drained and converted to buildable land. The upland between the ocean and the bay marshes was punctuated by low hills (e.g., Nob Hill, Telegraph Hill, Russian Hill) affording fresh breezes and superb views to the fortunate inhabitants wealthy enough to live on them. The imposition of the typical American grid street system on the irregular and steep slopes of San Francisco fostered the city's most famous means of transportation, the cable car.

But equally renowned as the Golden Gate is the region's physical nemesis, the San Andreas Fault. This fault—the locus of the 1906 and 1989 earthquakes—forms the tectonic boundary between the Pacific Plate and the American Plate. It skirts the Pacific Coast north of the Bay Area, passes San Francisco just offshore and converges with the mainland a few miles south of the present city limits. It then extends inland bisecting the peninsula south of San Francisco and continuing southeasterly toward Southern California. (See Figure 8-2.)

The "East Bay" was originally a marshy shoreline, backed by an abrupt but not lofty range of coastal hills. The westward toe of these hills is defined by the Hayward Fault, which extends north and south through the East Bay cities. A branch of the San Andreas system, the Hayward is "an offspring even more dangerous than its treacherous parent."[5] The next major Bay Area earthquake is expected to be centered there.

Originally cloaked in grasslands with pockets of oak, the East Bay Hills blocked expansion of settlement beyond the bay shore during the nineteenth century. With the advent of the automobile and technology for building on steep grades, the west-facing slopes of the hills, with spectacular views of the bay and distant San Francisco, have been elite residential districts since the 1920s. The intermixture of flammable buildings, lush ornamental vegetation, and relict grasslands poses a chronic risk of wildfire at times of drought. Conversely, at times of heavy winter rains, the streams that dissect the hills become raging torrents accompanied by the risk of mudslides that crush or sweep away anything in their path.[6]

Settlement and Growth

European settlement first reached the Golden Gate in 1776 when the Mission San Francisco de Assisi was founded by Spanish Franciscan monks and military authorities. The mission and the Presidio (later a U.S. Army post until the 1990s) formed the nucleus of the future city. The United States gained sove eignty over California from Mexico in 1846. Prior to that time, American involvement on the California coast was largely through New England-based ships engaged in coastal trading, in competition with Russian vessels as described in the literary classic *Two Years Before the Mast* by Richard Henry Dana.[7]

San Francisco was quickly transformed, first by political annexation to the United States, and then by the discovery of gold in the Sierra Nevada a hundred miles east of the city. The city served the mining and agricultural regions with banks, warehouses, retail stores, and hotels. The arrival of the railroad from the East in the 1860s (terminating at Oakland) provided the final linkage that established the Bay Area as the nation's primary West Coast economic hub.

As early as the 1850s, San Francisco began to emerge as a city of wealth, philanthropy, gracious architecture, and culture. In 1866, the New York landscape architect Frederick Law Olmsted, Sr., was invited to propose the plan for what would become Golden Gate Park, the San Francisco counterpart to his celebrated design for New York's Central Park. In 1875, the Palace Hotel, financed with Comstock Lode mining profits, opened as what many considered the finest hotel in the world until its destruction in the 1906 earthquake and fire. In 1905, Chicago architect Daniel H. Burnham unveiled a "City Beautiful" plan for the city, commissioned by the Association for the Improvement and Adornment of San Francisco. This plan was adopted by the city a month after the 1906 earthquake and fire, but much of it would never be achieved.

San Francisco flourished in its dual role as the West Coast hub of commerce, finance, and government and as a key entrepot in the network of Pacific trade. In 1903, on the eve of America's worst urban disaster, a prescient observer foresaw the city's future role on the Pacific Rim:

The great triangle of the Pacific is destined to have its lines drawn between Hong Kong, Sydney, and San Francisco. Of these three ports, Hong Kong will have China behind it, Sydney, Europe, and San Francisco, America; and with America for a backing, San Francisco can challenge the world in the strife for commercial supremacy.[8]

THE 1906 EARTHQUAKE AND ITS AFTERMATH

By the turn of the century, San Francisco had a population of one-third of a million, and was one of the nation's leading cities. Other cities, most recently Chicago, had experienced catastrophic fires. But no city in North America had been consumed by fire triggered by an earthquake. Like the sinking of the *Titanic* in 1912, the San Francisco earthquake and fire of 1906 shattered conventional assumptions regarding the ability of "modern technology" to overcome natural perils. And just as the loss of the *Titanic* led to tightening of safety requirements for ocean vessels, the San Francisco earthquake would stimulate actions to reduce the effects of future disasters.

The earthquake occurred at 5:12 A.M. on Sunday morning, April 18, with a tremor lasting 40 seconds. A major aftershock occurred 13 minutes later followed by many others of lesser magnitude. The early morning hour of the disaster undoubtedly saved many lives, particularly in the downtown business district where falling masonry would have been lethal to daytime crowds. On the other hand, the length of shaking proved catastrophic to much of the city's hastily built housing stock and many victims were trapped in their beds under debris. Many structures collapsed, others leaned, and thousands of masonry chimneys collapsed onto roofs, in some cases starting fires (Figure 8-1). Fissures opened in pavements miles from the actual fault (the epicenter lay under the ocean just offshore). Shaking of ground and buildings was accompanied by a cacophony of fearful noises. Hundreds of church bells pealed as the earthquake rocked the city, waking the citizenry in terror.[9] The earthquake also struck many other communities especially in the 200-mile corridor between Salinas in the south and Fort Bragg in the north. In the Bay Area, San Jose and Palo Alto were badly damaged. Oakland and Berkeley were less affected since the Hayward Fault was not involved, and those cities rendered vital assistance to San Francisco.

The ensuing fires took some time to get underway. After the initial shock, the populace fled to the streets, unaware that most of the damage to their city was yet to occur. Unlike the conflagrations of London (1666), Chicago (1871), and Oakland (1991), wind was not a major factor nor was the region experiencing a drought. The primary cause for the burning of the city was the almost total failure of its water supply. As recently as October 1905, the National Board of

FIGURE 8-1 Damage to San Francisco City Hall and vicinity caused by the 1906 Earthquake and Fire. (California Historical Society FN=25157)

Underwriters had found San Francisco's water supply to be inadequate in capacity and water pressure to fight a major fire.[10] Whatever its capability, the earthquake promptly disabled the system by rupturing water mains and cracking storage tanks. As gas lines exploded and cooking fires set homes ablaze, the fire department could not put out even modest fires. Bay water was tapped to protect some structures within reach of available pumps (as was also done in the Mission District of San Francisco after the 1989 Loma Prieta quake). Some use was made of dynamite to clear a fire break in certain locations. But by and large, the firemen were helpless spectators as the city burned (just as their modern counterparts would be in Oakland in 1991).

The San Francisco Earthquake and Fire of 1906 was one of the world's worst urban conflagrations. It burned an area of 2800 acres or 490 blocks—six times the area burned by the Fire of London of 1666[11] and half again as much territory as the 1871 Chicago Fire. It killed approximately 500 people. The homes of three-fifths of the city's inhabitants and the entire business district were destroyed.

The 1906 catastrophe possessed many elements of a contemporary megacity natural disaster: (1) multiple interrelated hazards (earthquake and fire); (2) failure of lifelines (water, communications, transportation) that caused secondary impacts; (3) widespread structural damage due to inadequate building standards and prevalent use of wood for smaller structures; (4) resulting homelessness and joblessness of much of the working-class population; but (5) a nurturing external society that assisted in the immediate response to the disaster and to some extent in the longer-term recovery.

But a missing element in the 1906 catastrophe, as compared with late twentieth-century disasters, was any significant assistance from the national government. Financial aid totaling $9 million (1906 dollars) was provided by private contributions and other cities across the nation.[12] Carloads of food and supplies were sent by train to Oakland to be ferried across the Bay. The Red Cross at the request of President Theodore Roosevelt was entrusted with overall coordination of the relief effort. The U.S. Army assisted in keeping order and allocating supplies. International assistance totaling $474,000 was also contributed from 14 foreign nations (including $50 from Germany!). But the response was largely a spontaneous, "grass-roots" reaction to the widely publicized disaster. Federal disaster assistance and recovery policies did not appear in the United States until the 1950s.

The most important public action taken by the city to avoid a repetition of the disaster was the development of an external water supply. Within weeks after the event, the San Francisco officials proposed to dam the Hetch Hetchy River Valley in the Sierra Nevada, 150 miles east of the city. Since the site lay within Yosemite National Park, this proposal sparked a 10-year controversy as to whether such a project should be allowed on federal land. The issue for the first time pitted wilderness advocates, led by John Muir, against proponents of "wise use of natural resources," represented by Gifford Pinchot, director of the new U.S. Forest Service and advisor to President Theodore Roosevelt. In the end, the latter faction prevailed and construction of the dam began in 1913.[13]

The Hetch Hetchy Dam and reservoir have since provided San Francisco with a high quality source of drinking water. But the reliability of the system in the event of another major Bay Area earthquake remains in doubt. The Hetch Hetchy water pipes extend across the Hayward Fault in the East Bay and thus could rupture in an earthquake there, potentially leaving San Francisco once again deprived of its water supply. The East Bay Municipal Utility District, which provides water to East Bay cities, faces a similar threat: Its aqueducts from Pardee Reservoir just north of Hetch Hetchy also must cross the Hayward Fault.

San Francisco recovered rapidly from the 1906 catastrophe chiefly because the downtown business district was well-insured. Total insurance payments amounted to $5.44 billion (1992 dollars), by far the worst urban fire insurance loss in American history. (Next were the Chicago Fire of 1871, $1.96 billion, and the Oakland Firestorm of 1991, $1.75 billion.)[14] Reconstruction of the central city began immediately. By 1910, the city's population stood at 417,000, an

TABLE 8-1 Population Change in San Francisco Bay Area: 1970–1990 (in thousands)

	1970	1980	1990	1980–1990
San Francisco-Oakland-San Jose CMSA[a]	4,754	5,368	6,253	16.4
San Francisco PMSA[b]	1,482	1,489	1,604	7.7
Oakland PMSA	1,628	1,762	2,083	18.2
San Jose PMSA	1,085	1,296	1,498	15.6
Santa Cruz PMSA	124	188	230	22.1
Santa Rosa PMSA	205	300	388	29.5
Vallejo PMSA	251	334	451	34.9
City of San Francisco	716	679	724	6.9
City of Oakland	362	339	372	9.7

Source: Statistical Abstract of the United States, 1991, Table 36.

[a]CMSA (Consolidated Metropolitan Statistical Area).

[b]PMSA (Primary Metropolitan Statistical Area). (PMSAs are components of a CMSA. Both are defined by the U.S. Bureau of the Census.)

increase of 22 percent over 1900. Housing was being constructed rapidly to replace what was lost and to accommodate newcomers.

However, like London after its Great Fire of 1666, San Francisco declined to alter its basic pattern of streets and land use patterns as it rebuilt. It even ignored Daniel H. Burnham's "City Beautiful" plan for the redesign of the city that had been presented to civic leaders just before the fire.

> Few cities ever found themselves demolished, with a ready-made plan for a new and grander city already drawn up, awaiting implementation, and with money pouring in to help realize the plan. San Francisco chose to ignore its Burnham Plan, and decided instead to build at a rate and manner which made the city not only less beautiful than was possible, but more dangerous. The rubble of the 1906 disaster was pushed into the Bay; buildings were built on it. Those buildings will be among the most vulnerable when the next earthquake comes.[15]

This statement was prophetic: the city's Marina District, built on 1906 rubble, was badly damaged in 1989.

THE 1989 LOMA PRIETA EARTHQUAKE

The continued growth of San Francisco and the Bay Area was unscathed by natural disasters for 83 years after the 1906 disaster. During this period, the relative dominance of San Francisco diminished as urban development flourished in

TABLE 8-2 Probability of One or More Large Earthquakes on the San Andreas Fault System, as Estimated in 1988

	Expected Magnitude	Next 5 years	10 years	20 years	30 years
S. F. Bay Area	7.0	0.1	0.2	0.3	0.5
Southern San Andreas	7.5–8.0	0.1	0.2	0.4	0.6
San Jacinto Fault	6.5–7.0	0.1	0.2	0.3	0.5

Source: U.S.G.S. Working Group, 1988. Probability of Large Earthquakes Occurring on the San Andreas Fault. Open File Rpt. 88-398. Washington, D.C.: U.S. Government Printing Office.

other parts of the Bay Area, facilitated by the building of the Golden Gate and Oakland Bay bridges during the 1930s. Decentralization was further encouraged after World War II by federal guarantees for new home construction in outlying areas, by new freeways to reach those homes, and in the 1970s by the Bay Area Rapid Transit (BART) system that links East Bay cities with each other and with San Francisco. The East Bay (Oakland PMSA) now exceeds San Francisco in both absolute population and growth rate (Table 8-1).

During the long period of deceptive quiet following the 1906 earthquake, another major earthquake on the San Andreas Fault was awaited by scientists. In 1988, a U.S. Geological Survey Working Group on California Earthquake Probabilities (1988) estimated that a magnitude 7.0 earthquake would occur on the northern San Andreas (Bay Area) with a probability of 10 percent within five years and 50 percent within 30 years (Table 8-2).

Concurrently, the U.S. Geological Survey and the California Office of Emergency Services (OES) were closely monitoring seismic activity in the region. Two small earthquakes in the Loma Prieta area in June 1988 and August 1989 persuaded OES to issue advisories to local governments of an impending larger earthquake. This represented "the most significant use of earthquake forecasting for public policy purposes since the 1975 Haicheng earthquake prediction in the People's Republic of China."[16]

These forecasts were soon fulfilled: The San Andreas Fault reawakened with a 7.1-magnitude earthquake at 5:04 P.M. on October 17, 1989. The epicenter was located about 70 miles south-southeast of San Francisco, near the town of Loma Prieta (Figure 8-2). The fault ruptured over a distance of 25 miles at a depth of 12 miles below the surface; no surface faulting appeared.[17] Seismic shaking lasted 15 seconds, followed by numerous aftershocks. The event was felt southward as far as Los Angeles and northward to the Oregon border. It caused 62 known deaths, 3757 injuries, and left 12,000 people homeless. It caused over $6 billion in property damage and disrupted public transportation, utilities, and communications.[18]

The earthquake coincided with the beginning of the third game of the baseball World Series, which ironically matched two teams from the Bay Area, the

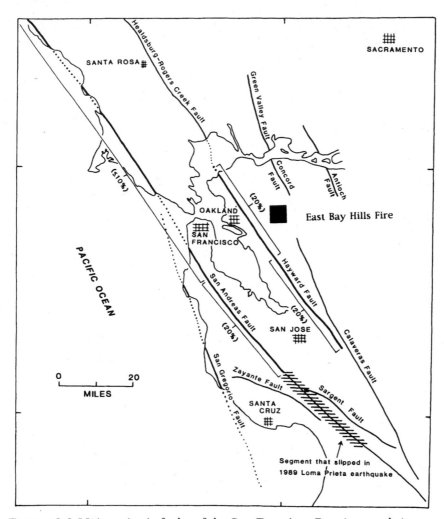

FIGURE 8-2 Major seismic faults of the San Francisco Bay Area and site of the East Bay Hills Fire of 1991. (Federal Emergency Management Agency/USGS)

San Francisco Giants and the Oakland Athletics. Sports commentators quickly shifted from the game preparations to the unfolding spectacle of disaster. It was an unparalleled opportunity to raise public awareness of seismic hazard. Warning of greater disasters to come was trumpeted by the news media around the world, as on the cover of *Newsweek* (October 30, 1989): "Bracing for the Big One: The Lessons of the San Francisco Earthquake."

Infrastructure Damage

Although the epicenter was in a rural upland well south of the major population centers of the Bay Area, it nevertheless caused dramatic and costly damage to infrastructure and older private buildings on both sides of the Bay (Figure 8-3). The most horrifying result of the earthquake was the collapse of a 1.5-mile double-deck section of a major commuter freeway constructed on bay mud in Oakland. The collapse trapped hundreds of cars and accounted for 41 of the 62 deaths attributable to the earthquake (Figure 8-4). For several days, the nation witnessed on television the heroic rescue of survivors from their vehicles trapped under the collapsed rubble of the upper deck. The last living victim was removed from his car after 90 hours.[19] (Even more extensive freeway damage would occur in the January 17, 1994, Northridge Earthquake in Southern California.)

Another lifeline failure was the collapse of a 50-foot section of the upper deck of the San Francisco–Oakland Bay Bridge, the only direct vehicular link between

FIGURE 8-3 Damage from Loma Prieta Earthquake, October 17, 1989. The first story of this three-story building in the Marina District was damaged because of liquefaction; the second story collapsed. What is seen is the third story. San Francisco County, California. (G. Plafker. U.S. Geological Survey Circular 1045, Fig. 23B)

FIGURE 8-4 Damage from Loma Prieta Earthquake, October 17, 1989. Collapsed Cypress structure on Interstate 880. The second deck collapsed onto the first deck. Alameda County, California. (G. Plafker. U.S. Geological Survey Circular 1045, Fig. 30)

San Francisco and the East Bay. Its closure for repairs forced the cross-bay commuters to take circuitous routes far to the north or south or to rely on the BART rapid transit system, which was not damaged. Cross-bay ferry service was also revived. In San Francisco, Interstate Highway 280 and the incomplete Embarcadero Freeway were also damaged. (The latter was subsequently scheduled for demolition.) Elsewhere, many local roads and highways were blocked by landslides triggered by ground shaking.

Structural damage was concentrated in cities close to the epicenter, particularly Santa Cruz. But many older unreinforced masonry buildings were damaged in San Francisco and Oakland, including its city hall. Damage in both cities was primarily caused by ground shaking on unconsolidated filled lands along the Bay. As in 1906, local water mains ruptured, leading to loss of water pressure for firefighting. A fireboat helped to pump bay water and local citizens organized bucket brigades to save the Marina District.

Elsewhere, damage was more related to the intensity of ground shaking than to distance from the epicenter.[20] Many critical facilities including several hospitals were damaged in locations where hazardous soil conditions could have been recognized. Stanford University in Palo Alto, which was severely hit by the 1906 earthquake, suffered $160 million in damage in Loma Prieta. Public schools were generally not badly damaged due to earthquake construction codes adopted after the 1933 Long Beach Earthquake in Southern California. The East Bay Municipal Utility District (EBMUD), which supplies water to 1.5 million people in the East Bay region, experienced some 200 local water main breaks, but its main supplies were not affected.[21]

In general, the earthquake caused four kinds of impacts from infrastructure damage:[22]

1. Direct physical and economic damage to the systems themselves
2. Diminished ability to carry out emergency response activities
3. Inconvenience due to temporary service interruption
4. Longer-term economic losses due to limits on recovery

Emergency Response

Preparation for a major earthquake in either the Bay Area or Southern California has been underway for many years. A megacity disaster of the expected magnitude is a national and even an international catastrophe (as in its impact on the world insurance industry). Planning involves not only local, county, and regional authorities but also the state of California and the Federal Emergency Management Agency (FEMA) and other federal agencies. Only three months before Loma Prieta, a training exercise ("Response '89") involving a hypothetical 7.5-magnitude earthquake on the Hayward Fault tested response capabilities in the Bay Area. After the real earthquake, FEMA reported that emergency responders—both governmental and nongovernmental—functioned smoothly but resources were stretched to the limit: "The entire system of emergency responders had reached a total saturation point, and if the disaster had been of any larger proportions, the total system would have broken down."[23]

Emergency response to Loma Prieta was hindered by many adversities to be expected in such a disaster, such as overload of telephone circuits, blockage of highways, rupture of gas lines, and temporary loss of water pressure and electrical power. On the other hand, several factors helped to reduce potential losses. The epicenter was relatively remote from the Bay Area's major population centers. Schools were not in session. Winds were calm, which assisted in the containment of fires. Commuter traffic was less than normal for 5 P.M. due to the World Series game. Many key personnel were still at their workplaces and available to staff emergency operations centers within minutes after the earthquake.

Property Damage and Individual Mitigation

The earthquake damaged more than 22,000 residential structures, 1567 commercial buildings, and 137 public buildings.[24] A total of 3957 structures were "red tagged" by local building inspectors as "uninhabitable." Most of the damage affected older buildings constructed before the state's Uniform Building Code was updated in 1973 (and revised further in 1988) to reflect seismic risk. Unreinforced masonry construction had long been prohibited for new buildings, thus reducing the risk of total building collapse as occurred in recent earthquakes in Mexico City, Armenia, and elsewhere. Modern office towers in downtown San Francisco swayed with ground motion, as they were designed to do. The preventive removal of projecting masonry and ornamentation from most older buildings also helped to avoid showers of rubble on people in the streets below[25] although 12 people died from this cause.

But damage to neighborhoods, homes, and personal property was widely experienced throughout the southern and central parts of the Bay Area. Some 22,000 residential structures were damaged to some degree. Post-earthquake survey respondents in San Francisco and Santa Cruz counties reported damage as displayed in Table 8-3.[26] Approximately two-thirds of San Francisco respondents and 87 percent of those in Santa Cruz County reported slight to moderate dam-

TABLE 8-3 Types and Distribution of Damage Due
to Loma Prieta Earthquake Reported by Residents
Surveyed in San Francisco and Santa Cruz Counties

| Neighborhood | San Francisco | | Santa Cruz | |
Damage	Percent	Number	Percent	Number
None	14.7	108	2.0	18
Slight	48.1	353	13.0	119
Moderate	20.7	152	50.0	459
Severe	16.5	121	35.1	322
Residence Damage				
None	37.7	27	13.4	123
Slight	55.2	405	51.1	469
Moderate	6.4	47	31.7	291
Severe	0.7	5	3.8	35
Household Items				
None	46.9	344	4.7	43
Slight	43.1	316	34.9	320
Moderate	9.5	70	50.8	468
Severe	0.5	4	9.7	89

Source: Modified from P.W. O'Brien and D. S. Mileti, 1993. "Citizen Participation in Emergency Response," in P.A. Bolton, ed., *The Loma Prieta, California, Earthquake of October 17, 1989—Public Response.* U.S. Geological Survey Professional Paper 1553-B. Washington, D.C.: U.S. Government Printing Office, Table 3.

age to their homes. Damage in "the neighborhood" was reported by 85.3 percent of those contacted in San Francisco and by 98 percent in Santa Cruz.

This widespread incidence of personal damage experience contrasts sharply with the well-documented tendency of Californians to decline to purchase catastrophic earthquake insurance and to eschew most other types of personal hazard mitigation activities. Even direct experience with damage in the Loma Prieta Earthquake failed to motivate a substantial increase in the purchase of earthquake insurance.[27] In a post-earthquake follow-up survey of residents in four California counties who had been contacted before Loma Prieta, the incidence of earthquake insurance purchase was found to have increased only marginally, and inversely to distance from the epicenter (Table 8-4, first part).[28]

In 1990, the California legislature established a mandatory earthquake insurance program providing up to $15,000 in benefits funded out of a surcharge on household insurance policies that would assist victims of lower-level damage and provide at least part of the amount deducted from higher-level losses. Palm and Hodgson also found a comparable reluctance to undertake other personal mitigation activities involving any monetary cost (Table 8-4, second part).[29]

Some 12,000 persons were left temporarily homeless by the earthquake, many of them poor. As in other disasters, the Red Cross moved quickly to set up shelters and emergency feeding stations. About one week after the event, 44 shelters were in operation, sheltering approximately 2500 persons a night; the remainder of the 12,000 either were housed with friends or relatives or were living in tents, vehicles, or otherwise improvising.[30]

TABLE 8-4 Earthquake Mitigation Actions Taken Before (1989) and After (1990) Loma Prieta by Residents Surveyed in Four California Counties

Survey Counties	Percent with earthquake insurance before Loma Prieta	Percent increase in earthquake insurance after Loma Prieta
Santa Clara	40%	+11%
Contra Costa	22	+7
Los Angeles	40	+6
San Bernardino	34	+1
	Percent incurring expense for other EQ mitigation actions 1989	Percent of same, 1990
Santa Clara	9%	31%
Contra Costa	7.8	10.4
Los Angeles	5	3.3
San Bernardino	5	6

Source: Modified from Risa Palm and Michael E. Hodgson, 1992. *After a California Earthquake: Attitude and Behavior Change.* Geography Research Paper 233. Chicago: University of Chicago Press, Figure 6 and Table 13.

Federal Emergency Response and Mitigation

Unlike the situation in 1906, federal disaster assistance in the United States today is massively available pursuant to a presidential disaster declaration. Only two months before Loma Prieta, Hurricane Hugo had spread devastation across the Virgin Islands, Puerto Rico, and North and South Carolina, resulting in disaster declarations for 113 counties, most of them rural and isolated. FEMA, the lead agency for coordinating federal disaster response, was hampered by the vast geographic area affected and was criticized for reacting slowly to that disaster. By contrast, the Loma Prieta disaster involved disaster declarations for only 12 counties that were all within or adjacent to the Bay Area.

FEMA quickly opened 18 disaster application centers, including five mobile offices. By the end of January 1990, 77,654 individuals, families, and businesses had registered at these centers for federal assistance, including about 40,000 applicants for temporary housing and 31,255 applicants for individual and family grants of $11,000 per household. The Small Business Administration issued 34,976 home and personal property loans and 14,773 business loans. Federal assistance by then had amounted to about $202 million.[31] As of August 1993, total federal costs for the Loma Prieta disaster were projected to reach $689 million.[32]

The population of victims, like that of California in general, was highly diverse. Special problems concerning temporary housing arose in Watsonville and other localities, reflecting long-standing ethnic and class antagonisms: "The earthquake and the subsequent housing crisis highlighted the preexisting deficiencies and inequities in Watsonville housing and provided the circumstance in which a new political agenda could be formulated, with housing as its focus."[33] One dispute involved demands by Mexican-Americans that mobile homes provided as emergency housing by FEMA remain indefinitely, being more habitable than the substandard housing damaged by the earthquake. Ironically, the federal government is empowered to address social needs in an emergency context that it may ignore at other times.

Section 409 of the Stafford Act requires that state or local governments receiving disaster assistance "shall agree that the natural hazards in the areas in which the proceeds of the grants or loans are to be used shall be evaluated and appropriate actions shall be taken to mitigate such hazards, including safe land use and construction practices." Accordingly, FEMA and the California Office of Emergency Services convened a state/federal hazard mitigation survey team shortly after the earthquake. The team involved 60 participants representing 32 federal, state, regional, and private entities with earthquake preparedness responsibilities. The ensuing report,[34] published three months after the disaster, offered 57 recommendations under the following headings:

a. Hazard identification and monitoring (3 recommendations)
b. Land use planning and regulation (2)

   c. Repair and reconstruction (18)

   d. Response planning (29)

   e. Insurance (3)

   f. Funding (2)

Clearly, the vast majority of recommendations address repair/reconstruction and response planning. By contrast, "land use planning and regulation" generated only two broad recommendations: (1) encourage local jurisdictions to incorporate available knowledge about . . . ground failure problems . . . into their general plans, buildings codes, and zoning ordinances; and (2) encourage [state] legislation to delineate specific zones that are susceptible to liquefaction and slope failure. The first is too generalized to assess whether it has been implemented; the second has not been.

The Next Bay Area Earthquake

Once the dust settled, it was clear that Loma Prieta was not the long-awaited "Big One" comparable to the 1906 earthquake. A new working group was assembled by the U.S. Geological Survey to assess the risk of further large earthquakes in the region. This review sought (1) to reassess earlier seismic data in light of Loma Prieta; (2) to evaluate the effects of Loma Prieta on other faults in the Bay Area insofar as possible; and (3) to incorporate new data on slip rate and recurrence interval on the Hayward and Rodgers Creek faults. The resulting report estimated *a 67 percent probability of one or more large earthquakes in the Bay Area within the next 30 years,* a substantial increase over the 50 percent probability estimated in the 1988 Working Group report.[35]

The Association of Bay Area Governments in 1992 estimated the numbers of housing units likely to be rendered uninhabitable by a 7.0-magnitude earthquake under several geographic scenarios (Table 8-5). These estimates reflect

TABLE 8-5 Estimated Total Dwelling Units Rendered Uninhabitable by Selected Earthquake Scenarios

Fault Involved	No. of Uninhabitable Units
San Andreas: Peninsula	8,154
Hayward: north and south	57,045
Hayward: north segment only	39,789
Hayward: south segment only	27,777
Healdsburg: Rodgers Creek	10,516
Loma Prieta (as per model)	3,323
Loma Prieta 1989 (actual)	3,957

Source: Adapted from Association of Bay Area Governments, 1992. *Estimates of Uninhabitable Dwelling Units in Future Earthquakes Affecting the San Francisco Bay Region.* Oakland, Calif.: ABAG, Table 1.

only damage due to ground shaking and do not include possible further damage due to fires, landslides, and other causes. Nevertheless each scenario except for the San Andreas envisions far more dwelling units becoming uninhabitable than in Loma Prieta, up to 29 times more in the case of a combined north-south Hayward event. The costs to insurers, government, and homeowners would all be vastly increased. And the post-disaster issues of emergency response, sheltering, and rehousing would be overwhelming.

The next major Bay Area earthquake most likely will occur on the Hayward Fault in the heavily urbanized East Bay. The USGS Working Group estimated a risk of nearly 30 percent that an earthquake of at least magnitude 7.0 will occur on the southern or northern segments of the Hayward Fault within the next 30 years. The last earthquakes on those segments occurred in 1836 and 1868 respectively. The Hayward Fault is "probably the most built-on fault in the world."[36] About 1.2 million people live within the epicentral region[37] of a potential 7.0 tremor on the Hayward Fault, 10 times the population of 130,000 within the comparable area affected by Loma Prieta (USGS Working Group 1990, 4). *Ground shaking in the East Bay would be at least 12 times worse in a 7.0-magnitude Hayward Fault earthquake than occurred there due to Loma Prieta.*[38] It is thus likely that the East Bay, which survived the 1906 and Loma Prieta events with relatively light damage aside from the expressway and bridge failures, will be the vortex of the next major Bay Area earthquake. Loma Prieta, which stretched emergency response to the limits, may in retrospect be viewed as merely a dress rehearsal.

THE EAST BAY HILLS FIRE OF 1991

Two years after Loma Prieta, disaster again struck the East Bay, this time in the form of wildfire in the home-strewn hills overlooking Oakland and Berkeley. The area affected was compact and the victims relatively affluent, as compared with the earthquake. But the Oakland–Berkeley Hills Fire of October 20, 1991, reminded disaster planners worldwide that natural hazards may be vastly aggravated by unwise building patterns and neglect of common-sense hazard mitigation (Figure 8-5).

Oakland and Berkeley extend from the filled flatlands along the bay shore to the steep, west-facing foothills of the Coastal Range. The combination of bay views, convenient access to urban pursuits, and semi-wild surroundings has long attracted the wealthy and talented to live in the East Bay Hills. During the 1920s and 1930s, the hills were interlaced with narrow winding roads lined with thousands of medium-sized, California-style homes on lots as small as one-eighth of an acre (5000 square feet).

The East Bay Hills region is exceptionally hazardous even by California standards. The Hayward Fault extends north and south along the entire length of the

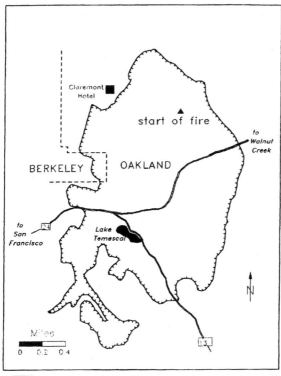

FIGURE 8-5 Approximate area burned by the Oakland/Berkeley Hills Fire, October 20, 1991. (Source: Matrix Maps)

foothills, dividing the flatlands from the slopes. Many of the area's public facilities, including the football stadium of the University of California at Berkeley, lie directly over the fault. As discussed earlier, the Hayward Fault has not noticeably shifted since 1868 and is estimated to have a 30 percent chance of a major earthquake in the next 30 years. Landslides, flash flooding, and debris flows are additional physical hazards in the East Bay Hills.

The risk of urban-wildland fire in this region and similar areas in the West arises from several interrelated factors:

- *Vegetation.* The native oak and grassland ecosystem has been transformed by the introduction of California landscape trees, especially eucalyptus and Monterey pine. These species contain much resin and are highly flammable in comparison with oaks. Furthermore, dead biomass from the eucalyptus accumulates on the ground unless it is removed, providing an excellent source of fuel.

- *Fire Suppression.* The presence of homes has motivated control of fires that occur naturally, causing an increase in fire hazard due to accumulation of dead vegetation in remaining grasslands (as within regional parklands).

- *Drought.* Vegetation in 1991 was desiccated from a 5-year drought.
- *Building Materials.* Most homes are constructed of wood with "shake" or cedar shingle roofs that are characteristic of California woodland architecture. Also most homes have wooden decks and other projections.
- *Narrow Roads.* The steep, winding roads serving the hills impede the mobility of police, fire, and medical vehicles and also obstruct evacuation by car. Abandoned vehicles may block such roads entirely.
- *Wind.* Late summer and autumn in California are marked by hot, dry winds known as "Santa Ana winds." High-pressure systems over the interior deserts of the Great Basin sometimes cause strong winds to blow westerly toward the coast. Steep terrain such as the East Bay Hills intensifies wind speeds and turbulence.
- *Water Supply.* Firefighting in the hills may be impaired due to rupture of water lines and loss of water pressure.
- *Arson.* Wildland fires in California are often thought to be ignited by arsonists.
- *Floods and Debris Flows.* Once a fire occurs, the burned areas may later experience flash floods, mudslides, and debris flows caused by heavy precipitation on exposed slopes.

California has a long history of major wildfires on the urban fringe (Table 8-6). Since 1923, at least 14 fires have occurred in the Oakland Hills, including seven in Wildcat Canyon where the 1991 fire began.[39] Today, about seven million people are estimated to live in regions of the state vulnerable to urban/wildland fires.[40]

There has been no lack of expert advice concerning urban-wildland fires in California. After the 1923 fire, a special commission report offered more than 100 recommendations for mitigating future losses, most of which were ignored.[41] After a series of Southern California fires, a 1980 Governor's Task Force on Chaparral Fire and Flood Risk Management offered another long list of recommendations concerning (1) risk reduction, land use, and conservation standards; (2) vegetation management and fire hazard reduction; and (3) fire command system improvements.[42]

Planning for natural disasters in urban areas must be conducted at a geographic scale commensurate with the problem. The fire hazard in the East Bay Hills was broader than the individual municipalities concerned, but narrower than the entire Bay Area. In 1982, a Blue Ribbon Fire Prevention Committee for the East Bay Hills was formed under the East Bay Regional Park District, which manages several large tracts of public open space in the hills. The district served as coordinator for a comprehensive review of the fire hazard affecting the entire Hills region. The resulting report[43] focused primarily on vegetation man-

TABLE 8-6 Major California Urban-Wildland Fires
1923–1993

Year	County	Acres Burned	Structures Burned
1923	Alameda	130	584
1961	Mariposa	41,200	106
1961	Los Angeles	6,090	484
1964	Napa-Sonoma	71,601	174
1964	Santa Barbara	61,000	94
1967	Riverside	48,639	61
1970	San Bernardino	53,100	54
1977	Santa Barbara	804	234
1978	Sonoma	11,504	64
1980	San Bernardino	5,482	65
1980	San Bernardino	41,472	355
1981	Napa	22,000	69
1982	Ventura/Los Angeles	57,000	65
1985	San Diego	200	64
1988	Nevada	33,500	312
1988	Shasta	7,800	58
1990	Santa Barbara	4,900	641
1991	Alameda (Oakland)	1,600	3,349
1993	Southern Calif.	272,000	1,200

Source: Adapted from California Governor's Office of Emergency Services, East Bay Hills Fire Operations Review Group, 1992. *The East Bay Hills Fire: A Multi-Agency Review of the October 1991 Fire in the Oakland-Berkeley Hills.* Sacramento: O.E.S. Annex A

agement and particularly the need to extend existing "fuel breaks" along the entire 25-mile length of the urban-wildland interface zone. It also proposed that a new special assessment district be created to manage vegetation throughout the area at risk. These recommendations, however, were not accomplished prior to the October 1991 fire.[44]

The Firestorm

The East Bay Hills Firestorm of October 20, 1991, was the third worst urban fire disaster in U.S. history after the San Francisco Earthquake of 1906 and the Chicago Fire of 1871. Although the area burned (about 1700 acres) was modest compared with other California fires, it was by far the worst urban-wildland fire in terms of structural damage. Twenty-five people died and over 150 were injured. It destroyed or damaged 2621 homes and 758 apartment and condominium units in about 9 hours. About 5100 people, many of them elderly, were left homeless. Damage amounted to at least $1.5 billion.[45]

The fire erupted under textbook conditions. Daytime temperatures hovered around 90 degrees F, and relative humidity was 17 percent. Hillside vegetation was bone dry. Dead plant material littered the ground, and trees overhung many homes despite the long drought and warnings of fire danger. Hot, dry Santa Ana winds blew from the east on the morning of the conflagration. The State Department of Forestry had issued a "Red Flag" warning of potential fire hazard, but few residents took notice.

It began on Saturday, October 19 with a small brush fire "of suspicious origin" high up Wildcat Canyon near the ridge line of the Hills in Oakland. The fire was apparently extinguished that day but was not watched overnight. On Sunday morning, with Santa Ana winds blowing at medium strength, the fire reignited and rapidly spread downhill, leaping from house to house. Fire personnel who returned to the scene were overwhelmed. About 790 houses burned in the first hour after it began to spread. Turbulent winds generated by the fire itself spewed burning material in all directions. The fire crossed an eight-lane highway (Route 24) and continued to consume homes and vegetation farther downslope. Public orders to evacuate were difficult to communicate in the absence of sirens.

The huge smoke plume itself was the primary warning to Hills residents. Many started packing their cars only to find it impossible to drive down the obstructed roads. Cars by the hundreds were abandoned and destroyed as individuals literally ran for their lives, leaving the roads impassable. There was little the residents could do at the last moment to avoid the impending disaster, other than saving themselves and whatever they could carry. Sixty years of building on the hills had created a hazard that no one could individually undo. It was a classic "tragedy of the commons." All were swept up in the common peril and personal consequences depended more on the fluke of the winds than on individual actions. (See the first quote at the beginning of this chapter.)

Emergency response was massive but ineffectual. The fire departments of Oakland and Berkeley were depleted by funding cuts and were neither well-trained nor equipped to fight a fast-moving urban/wildland fire. In response to their appeals for help, an armada of local, state, and federal fire fighters converged on the scene including an incredible 440 local engine companies, 6 air tankers, 16 helicopter units, 8 communications units, 2 management teams, and 88 fire chiefs.[46] Despite acts of great courage, firefighters were helpless. Communication was hampered by overload of radio and telephone circuits. Smoke and traffic congestion obstructed mobility. And Oakland hydrants were incompatible with the hose couplings of other municipal fire departments.

These problems, however, were overshadowed by the loss of pressure in the water distribution system within the Hills. This failure resulted from three factors: (1) all but one of eleven East Bay Municipal Utility District (EBMUD) reservoirs serving the area ran dry due to outflow through ruptured mains, melted pipes, and open faucets; (2) electrical power failed in the area as the fire burned transmission facilities, thus disabling water supply pumping stations; and (3) EBMUD lacked emergency generators to operate those disabled pumps.[47]

The conflagration was intensely hot and fast-moving. Temperatures at its center exceeded 2000 degrees F, hot enough to boil asphalt.[48] Firestorm conditions began to develop within 15 minutes of ignition of the first structure. A thermal inversion level formed at approximately 3000 feet that trapped heat from the fire and intensified preheating of vegetation and structures in the area. The fire consumed hundreds of structures simultaneously. Some houses were believed to have exploded from within, perhaps due to the intense heating of interior spaces by picture windows facing the flames (a form of greenhouse effect).[49] Propane and other flammable substances also contributed to the loss of many homes. Once ignited, there was little or no chance to save any home. All that remained the next day were concrete foundations, chimneys, and charred tree trunks. In addition to their homes, many fire victims lost personal mementos, pets, art and music collections, and the contents of computer disks. Many of these losses were not covered by fire insurance.

The flames reached within a few hundred yards of the opulent and historic Claremont Hotel. If the winds had continued, the fire could have consumed much of Berkeley, including the University of California campus. However, at about 7:00 P.M. on Sunday evening, the wind shifted and diminished in force, nudging the fire back over areas already burned. This allowed the firefighters to contain and ultimately control the fire.

The Recovery Process

On the morning after the disaster, the mayor of Oakland vowed that the Hills would be rebuilt. This reflected both solicitude for the victims and the reality that the burned area represented some of Oakland's highest property tax valuations, as well as clientele for many of the city's businesses and contractors (Figure 8-6).[50]

The option of acquiring the burned area and adding it to the existing system of regional parks in the Hills was never seriously considered. Public acquisition costs would have been sizeable and Oakland would have lost part of its tax base. Nevertheless, the failure to assess the economic and environmental impacts of not rebuilding represents a major shortcoming in the post-disaster recovery process. Federal assistance, to the contrary, expedited the rebuilding process, as described below.

In fact, the cost of acquisition might have been an economically feasible alternative to reconstruction. The average price of lots sold in the burned area as of October 1993 was approximately $130,000.[51] If the approximately 3000 vacant building lots could have been acquired for an average cost of $130,000 each, this

FIGURE 8-6 Rebuilding of Oakland Hills in progress, 1993. (Photo by R.H. Platt)

would amount to about $390 million plus administrative costs and lost tax revenues. As compared with the public costs of reconstruction and restoring public streets and utilities to the area (on the order of $100 million), this might not have been excessive. Emotional costs of those who rebuilt under difficult conditions would have been eased. Fire victims would have received both their fire insurance proceeds and their site acquisition payment. And the risk of future conflagrations, possibly triggered by a Hayward Fault earthquake, would have been greatly reduced.

But Oakland relies on the Hills as its highest residential tax-generating district and was not interested in a buyout, even if funded externally. Private owners and public officials alike, however, shared a pervasive determination to understand the causes of the disaster and, to the extent possible, to rebuild more safely in light of the fire and seismic risk. Both professional and journalistic post-disaster assessments were produced in abundance. Seldom has a natural disaster been so thoroughly and competently examined from a multitude of perspectives. One of the earliest of these was a special series by the *Oakland Tribune:* "Firestorm: Bitter Lessons." The disaster also spawned the *East Bay Phoenix Journal,* which served as a voice and forum for fire victims. An anthology of essays, poems, and photographs by fire victims was published privately with proceeds donated to the Alta Bates Burn Center.[52]

Unlike the Loma Prieta Earthquake, private insurance played a dominant role in financing the rebuilding of the Hills. One year after the disaster, 3954 claims amounting to $1.4 billion had been filed with 49 insurance companies, averaging more than $350,000 per household.[53] Many of these were disputed by the insurers as to the amount claimed, leading to frustration and often to lawsuits filed by claimants. However, as settlements have eventually been reached on most claims, the fire victims have received substantial payments. About two-thirds of the owners held guaranteed replacement cost coverage and also received alternative housing costs and several kinds of federal and state tax relief. The average net worth of families who lost their homes was estimated to rise from 10 to 25 percent compared with prefire levels.[54]

Reconstruction in the Hills was California's largest building boom in 1992 and 1993, creating some 11,000 construction jobs during a time of statewide recession. As of October 12, 1993, building permits had been issued for about half the homesites burned in Oakland (1368 out of 2777) and for over half those burned in Berkeley (39 out of 69). Homes completed in those communities numbered 434 and 20, respectively.[55] This represented over one billion dollars already invested in rebuilding with several years of construction still ahead.

Post-Disaster Assessments

Government agencies—city, county, special district, state, and federal—played many roles in the recovery from the fire. These included emergency response and rescue, infrastructure repair, grants and tax subsidies to the victims, administration of the rebuilding process, and research on the causes of the disaster. The occurrence of the disaster at the doorstep of the University of California at Berkeley spawned a variety of studies and public outreach efforts relating to the fire. Post-disaster reports spanned many disciplines including urban forestry, geology, urban planning, landscape design, fire fighting, mental health, and water supply (Table 8-7).

The federal government officially entered the recovery process under a disaster declaration issued by President Bush on October 22, 1991, for Alameda County. For the second time in two years, FEMA established a disaster center in Oakland and activated the many forms of aid available under the Stafford Act. The effects of the fire were felt more heavily on property and less on public infrastructure, in comparison with the earthquake. The number of structures destroyed or badly damaged was roughly comparable in the two disasters: 3957 were "redtagged" after the earthquake and about 3300 were burned in the fire. But as of late 1994, the total cost of federal assistance following the fire was $513 million, or about 10 percent of federal assistance by then committed to recovery from the earthquake.[56]

Approximately four-fifths of federal funding after the fire was spent on reimbursing state and local governments for their response and recovery costs. Part

TABLE 8-7 East Bay Hills Firestorm: Selected Post-Disaster Reports

- FEMA, *Hazard Mitigation Report for the East Bay Fire in the Oakland–Berkeley Hills* (Federal Disaster Declaration 919-DR-CA, 1992).
- California Governor's Office of Emergency Services, *The East Bay Hills Fire: A Multi-Agency Review of the October 1991 Fire in the Oakland/Berkeley Hills.* Report prepared by the East Bay Hills Fire Operations Review Group (February 27, 1992).
- Task Force on Emergency Preparedness and Community Restoration, *Final Report.* (February 3, 1992). The Task Force was established jointly by the mayors of Berkeley and Oakland and included committees on (1) Emergency Preparedness; (2) Communications; (3) Forestry and Vegetation; (4) Infrastructure and Development; and (5) Planning, Zoning, and Design.
- Ad Hoc Council on Replanting Needs (ACORN) (a project of the University-Oakland Metropolitan Forum), "Guide to Landscaping for Fire Safety," collection of materials issued March 1992, second edition issued October 1992.
- Kenneth Topping, *Report on Oakland Hills Fire Prevention and Suppression Benefit Assessment District.* October 15, 1992 (consultant report prepared for the city of Oakland).
- Hills Emergency Forum, *1993 Report.* Ongoing intergovernmental forum including (1) East Bay Fire Chiefs Consortium and (2) Vegetation Management Consortium.
- East Bay Municipal Utility District, *Firescape: Landscaping to Reduce Fire Hazard* (1992).
- East Bay Municipal Utility District, *Preliminary Report on EBMUD's Response to the East Bay Hills Firestorm* (March 24, 1993); *Phase I Final Report* (July 15, 1992); *Phase II Final Report* (January 21, 1993).
- National Fire Protection Association, *The Oakland–Berkeley Hills Fire, October 20, 1991* (1992).

of this federal aid funded a Community Restoration Development Center (CRDC) as a field office of the Oakland city government. Situated in a former supermarket close to the fire scene, the CRDC provided a "one-stop" facility where fire victims could meet with representatives of city, state, and federal agencies, utilities, insurance companies, medical providers, and social services. Planning staff at the CRDC reviewed proposed rebuilding plans. A computerized geographic information system (GIS) covering ownership, site characteristics, vegetation, utilities, and other data for each parcel in the burned area expedited permit approval.[57]

As required by the Stafford Act, a hazard mitigation team convened by FEMA offered a variety of (increasingly familiar) recommendations concerning vegetation management, nonflammable roofs, widening of roads, fire codes, flood hazards, and other matters.[58] The California Governor's Office of Emergency Services in a parallel report reviewed emergency response issues, particularly communications and fire fighting. Among other proposals, it urged the adoption of compatible hydrant connectors among communities to facilitate mutual response to fires, and the installation of a permanent auxiliary pumping system for refilling all upper water zones in hill areas.[59]

An immediate concern following urban-wildland fires in California is the risk of landslides, debris flows, and erosion on steep slopes with the advent of the winter rainy season. The city of Oakland spent $5 million of federal money on aerial reseeding of the burned area and placing hundreds of straw bale check dams in gullies and channels.[60]

The most comprehensive policy review of the Oakland fire was conducted by a regional Task Force on Emergency Preparedness and Community Restoration created by the mayors of Oakland and Berkeley a week after the fire. The task force was asked to prepare a detailed report within eight weeks to guide the recovery of the burned area and protect other areas in the East Bay Hills. With membership that included city and county officials, university faculty, utilities, local businesses, and fire victims, the task force included five work groups:

Emergency Preparedness to evaluate the emergency response systems of the cities, including fire stations, communications, early warning systems, and the use of volunteers

Communications to improve media and public information during a disaster

Forestry and Vegetation to develop landscape standards and other guidelines that reduce the potential for fire to spread

Infrastructure and Development to increase the quality of street systems, utilities (water, electricity, and telephone) for the fire areas

Planning, Zoning, and Design to develop new zoning and design standards for the area to be rebuilt as well as for all potential fire areas

Among the nearly 300 recommendations developed by the work groups, the key proposals and their outcomes are displayed in Table 8-8.

Collaboration between city and university was especially fruitful regarding forestry and landscaping issues.[61] A grant from the San Francisco Foundation funded a city-university project named the "Ad Hoc Council on Replanting Needs" (ACORN) to educate property owners on fire resistant landscaping methods. Reinforcing this effort, the East Bay Municipal Utility District prepared a public information document entitled "Firescape: Landscaping to Reduce Fire Hazard," which it distributed as a public service to all of its water customers in the Hills.

Translating public education into political action, the voters of Oakland in 1993 approved an Oakland Hills Fire Prevention and Suppression Benefit Assessment District. The district overlays all hill areas of Oakland, whether or not burned in the 1991 fire. Each of the approximately 20,000 parcels of real estate in the district will be assessed $75 per year initially, which could rise to a maximum of $300 per year. This revenue will be used primarily for reduction of

TABLE 8-8 Task Force on Emergency Preparedness and Community
Restoration: Key Proposals and Status (1994)

- Improve local emergency services through a major bond issue for new radio communications and fire protection equipment. *Adopted*
- Increase the training and use of community volunteers to identify fire hazards and fight small fires. *Adopted*
- Require all homes in designated fire hazard zones to remove wood roofs, use prescribed landscaping methods and materials, and incorporate fire control methods. *In Progress*
- Establish a fire protection district for the fire danger zone from funds allocated for fire prevention activities. *Adopted*
- Move public utilities underground in the fire area, *Accomplished* and improve water supply to the area. *In Progress*
- Revise building codes to require the use of sprinklers *Not Adopted* and/or more fire-resistant building materials. *Adopted*
- Improve street access for firefighting equipment, *Not Adopted* and relocate some fire equipment to fire-prone areas. *Adopted*
- Limit the density of homes in fire-prone areas. *Not Adopted*

Source: Task Force on Emergency Preparedness and Community Restoration 1992. *Final Report* (February 3), mimeo. Status based on estimates by Rutherford H. Platt.

hazardous fuel loadings on public and private lands through various means (goat herds have been employed to graze public lands within the district). The program will fund inspections on private lands to ascertain violations of the Uniform Fire Code and will strengthen the capability of Oakland to combat future urban-wildland fires.[62]

The Task Force Committee on Planning, Zoning and Design addressed the controversial issues of how the Hills should be rebuilt. It was firmly committed to rapid rebuilding and gave little apparent consideration to the alternative of buying the burned areas for public open space or limiting density through zoning regulations. Despite the minuscule size of lots in the burned area—many are only 5000 square feet—the committee urged approval for reconstruction of homes within the same "footprint." It called for utilities to be underground and the adoption of "innovative parking solutions" to free up narrow roadways. (It did not, however, urge the widening of roads.)

Fourteen months after the fire, Oakland finally adopted a new overlay zone (S-14) for the burned area. It was quite generous to property owners, permitting enlargement of burned structures by 10 percent and exempting any plans submitted before its date of effectiveness regardless of the size of structures proposed. The S-14 zone establishes a variety of rules concerning minimum lot setbacks and sideyards, heights of buildings, parking, and landscaping. These rules are specified in relation to the particular slope and street frontage characteristics of the lot in question.

As of late 1996, the physical results of the ordinance were visible on the hillsides of Oakland: hundreds of very large, free-standing homes of eclectic design, often standing within 10 to 20 feet of each other (Figure 8-7). Fire hazard is reduced through the use of noncombustible roof materials, undergrounding of utilities, and limits on flammable vegetation. Off-street parking will be more abundant under the new zoning law.

But the roads are generally as narrow as before and larger homes may well produce more vehicles per household. And the sport utility vehicle trend suggests that vehicles themselves are becoming larger. Congestion of streets as a public safety hazard has therefore not been reduced in many areas. The proximity of the homes, the lack of internal sprinklers to deal with common house fires, and the likelihood of street congestion all create a sense of unease in light of the potential risk of a Hayward Fault earthquake in the next few decades. With some justification, private homeowners may assume that if it were not safe to rebuild, the government would prevent them from doing so.

FIGURE 8-7 Rebuilt neighborhood in Oakland Hills, 1996. Wood roofs are prohibited and vegetation is controlled, but the area lies just upslope from the Hayward Fault. (Photo by R.H. Platt)

CONCLUSION

The Bay Area and especially the East Bay region have been sorely tested by earthquake, wind, and fire since 1989. The Loma Prieta Earthquake and the East Bay Hills Fire each demonstrated the vulnerability of urban regions to lifeline failure—especially water, transportation, electricity, and medical facilities and systems. On the other hand, each displayed the beneficial effects of a well-organized "larger community" extending from the president of the United States to the professional staff of the local and regional agencies to whom private citizens and local city officials look for help in an emergency.

There is little cause for comfort in the fact that Loma Prieta and the fire were not more catastrophic. As reported by FEMA, the emergency response capabilities of the entire federal, state, regional, and local governments were stretched to the breaking point by the earthquake. Differences in a few key variables—proximity of the epicenter to San Francisco, magnitude of the quake, wind conditions, time of day—could have swamped the emergency response system. Similarly, if the Santa Ana winds had continued to blow, the 1991 firestorm could have engulfed the world-famous Claremont Hotel, the University of California Berkeley campus, and untold numbers of Berkeley and Oakland homes and businesses. Indeed, the Southern California firestorms of 1993 burned a far larger area (250,000 acres) although fortunately fewer structures (1200).

The juxtaposition of two unrelated natural disasters in the same urban region within two years, however, may be viewed as an unparalleled training exercise in disaster response and recovery. The following are some concluding observations.

The Regional Scale

Some American metropolitan areas have established regionwide hazard management agencies, such as the Denver Urban Drainage and Flood Control District.[63] Often, however, metropolitan areas are too large and individual cities are too small to serve as optimal geographic scales for disaster planning, response, and recovery. Urban regions need to be divisible into smaller subareas consistent with the functional requirements of particular services. Although the city of Oakland was a primary urban victim of both the 1989 earthquake and the 1991 firestorm, it could not, and did not, face these trials in isolation. The "East Bay" provided a psychic and functional region within which Oakland could be helped to recover.

Within urban regions, there is a high degree of interaction among government, corporate, nonprofit, and individual participants in addressing common needs. Long before Loma Prieta, there had evolved a strong tradition in California of mutual aid and intergovernmental cooperation. This is reflected in myriad intergovernmental agreements on public services, fire protection, and shared facilities. It also appears in strong and flexible institutions of regional governance. In the case of the East Bay, two key regional agencies are the East Bay Municipal Utility District and the East Bay Regional Parks District. Other institutions crucial to the disaster recovery process have included the University of California at Berkeley, the Association of Bay Area Governments, Pacific Gas and Electric, and Pacific Bell. Professional staff from these and other entities interact continuously on task forces, committees, and special projects such as the ACORN project of the University-Oakland Metropolitan Forum. Governmental, corporate, and philanthropic efforts are blended in an ongoing dialogue concerned with the welfare of the East Bay.

Post-Disaster Hazard Assessment

Both Bay Area disasters were followed by a deluge of post-disaster assessments prepared under federal, state, regional, local, and private auspices. Some of these reports, like the FEMA hazard mitigation report, were intergovernmental and interdisciplinary, but necessarily general in tone. Others were more focused on particular agency or function, such as urban forestry, soil stabilization, fire fighting, and medical care.

For post-disaster assessments to be useful, they must be (1) timely, (2) scientifically accurate, (3) prescriptive, and (4) practical. But they must not be overly constrained by political considerations. Members of official post-disaster assessment teams should be charged to "tell it like it is" and make whatever recommendations seem desirable without fear of criticism for suggesting politically unpopular approaches.

In the case of the East Bay Fire, the post-disaster assessments were professionally competent but politically cautious. The most prominent of the nonfederal assessments, by the Task Force on Emergency Preparedness and Community Restoration, articulated dozens of sound hazard mitigation measures. But it left unmentioned the ultimate mitigation measure—not rebuilding the Hills, or at least reducing density and widening streets. The public and political saliency of those options was perhaps slight, but they should have been considered. Public buyout of significant portions of communities has occurred in the case of Rapid City, South Dakota, after its 1972 flash flood, and at Times Beach, Missouri, after it was contaminated with dioxin in the 1970s. The Mississippi Valley Floods of 1993 prompted federal purchase of some 12,000 flooded properties. (See Chapter 7.)

Even more modest efforts to change land use and building patterns in the burned area after the fire were ineffectual. Streets remain as narrow as before and homes have become larger, albeit roofed with tile rather than shingle. Most of the impact of post-disaster assessments has been to improve public sector services with little or nothing expected of private owners. The specter of a Hayward Fault earthquake apparently played little role in the deliberations of the Task Force on Emergency Preparedness and Community Restoration.

Lifelines

Much has been said in this chapter about lifeline vulnerability in the 1906 Earthquake, the Loma Prieta Earthquake, and the East Bay Hills Fire. At the risk of overstating the case, there is no more significant issue for natural hazard planning in urban areas than lifeline reliability. California has achieved measurable improvement in the seismic survivability of individual structures due to building code improvements. Also, individuals and households are better able to protect themselves during a disaster due to public education programs. But such improvements in individual safety are effectively nullified by the increased vulnerability of public and private lifelines to failure due to natural disaster. The Northridge Earthquake of January 17, 1994, and the Kobe, Japan, Earthquake one year later demonstrated the vulnerability of lifelines.

After twice being found lacking, the East Bay Municipal Utility District has improved the resilience of its water supply system against earthquake and fire. Less subject to upgrading, the Bay Area's aging transportation arteries remain vulnerable to the kinds of failures that occurred in Loma Prieta and later in the Northridge Earthquake in the Los Angeles area. Larger earthquakes would vastly magnify damage to vital lifelines such as electrical services, dams, pipelines, and transportation.

The Unexpected

Perhaps the most daunting challenge for disaster managers is to plan for the unexpected. The interconnectedness of modern urban regions triggers cascading series of failures, not necessarily catastrophic individually, but cumulatively producing possible chaos, panic, and misery, as well as economic hardship. Dealing with the unexpected requires trained professionals equipped with reliable technology for damage assessment, decision making, and communication. Disaster response is a professional skill and an art. Perhaps the only silver lining in enduring a disaster is that the ability of the affected state or city to prepare for the next one is enhanced. Short of firsthand experience, the need to learn from disaster experience elsewhere is obvious. The San Francisco Bay Area has much to teach.

NOTES

1. Portions of Chapter 8 are derived from the author's chapter "Natural Hazards of the San Francisco Bay Region: Trial by Earthquake, Wind, and Fire," in James K. Mitchell, ed. *Crucibles of Hazard: Megacities and Disasters in Transition.* Tokyo: United Nations University Press (1999) by permission of the UNU Press.

2. San Francisco accounts for only 740,000 people or about 11 percent of the total CMSA population of 6.5 million. U.S. Bureau of the Census (1994). *Statistical Abstract of the U.S.* Washington, D.C.: U.S. Government Printing Office.

3. Risa Palm and Michael E. Hodgson (1992). *After a California Earthquake: Attitude and Behavior Change.* Geography Research Paper 233. Chicago: University of Chicago Press.

4. H.H. Bancroft (1888). *History of California.* San Francisco: The History Company, as quoted in James E. Vance, Jr. (1964). *Geography and Urban Evolution in the San Francisco Bay Area.* Berkeley: Institute of Governmental Studies, University of California, 4–5.

5. Peter I. Yanev (1991). *Peace of Mind in Earthquake Country.* San Francisco: Chronicle Books, 36.

6. The cycle of drought, fire, and flood is best described by John McPhee (1989). "Los Angeles Against the Mountains," in *The Control of Nature.* New York: Farrar, Straus & Giroux. Also see Brenda Bell (1999). "The Liquid Earth." *The Atlantic Monthly* (January), 58–71.

7. Richard Henry Dana (1832/1964). *Two Years Before the Mast,* 2 vols. Boston: Little, Brown.

8. C.A. Keeler (1903). *San Francisco and Thereabout.* San Francisco: The California Promotion Committee, 94.

9. William Bronson (1986). *The Earth Shook, the Sky Burned.* San Francisco: Chronicle Books.

10. Ibid.

11. Immediately after the 1666 Fire of London, King Charles II convened a royal commission that included the architect Christopher Wren, to investigate the causes of the disaster. The commission recommended specific legal reforms including new building regulations and a procedure for their enforcement. The recommendations were adopted into law by Parliament in the Act for Rebuilding London of 1667. See Rutherford H. Platt (1996). *Land Use and Society.* Washington, D.C.: Island Press, Ch. 3.

12. Bronson, note 9.

13. Roderick Nash (1982). *Wilderness and the American Mind.* New Haven: Yale University Press.

14. *The San Francisco Chronicle,* October 31, 1993, B1.

15. Thomas, Gordon, and Max Morgan Witts (1971). *The San Francisco Earthquake.* New York: Stein and Day, 274.

16. Federal Emergency Management Agency (1990). *Hazard Mitigation Opportunities for California.* State-Federal Hazard Mitigation Survey Team Report for Disaster Declaration FEMA 845-DR-CA. San Francisco: FEMA.

17. R. Bolin (1993). "The Loma Prieta Earthquake: An Overview," in R. Bolin, ed., *The Loma Prieta Earthquake: Studies of Short-Term Impact.* Program on Environment and Behavior Monograph No. 50. Boulder: University of Colorado Institute of Behavioral Science.

18. G. Plafker and J.P. Galloway, eds. (1989). *Lessons Learned from the Loma Prieta, California, Earthquake of October 17, 1989.* U.S. Geological Survey Circular 1045. Washington, D.C.: U.S. Government Printing Office.

19. *Newsweek,* October 30, 1989, 32.

20. FEMA (1990), note 16.

21. Ibid.

22. Stephen French (1990). " A Preliminary Assessment of Damage to Urban Infrastructure," in R. Bolin, ed., *The Loma Prieta Earthquake: Studies of Short-Term Impact.* Program on Environment and Behavior Monograph No. 50. Boulder: University of Colorado Institute of Behavioral Science.

23. Federal Emergency Management Agency (1991). *The Loma Prieta Earthquake: Emergency Response and Stabilization Study.* Contract report by the National Fire Protection Association. Washington, D.C.: U.S. Government Printing Office, 40.

24. Ibid., 21.

25. Plafker and Galloway, note 18.

26. P.W. O'Brien and D.S. Mileti (1993). "Citizen Participation in Emergency Response," in P.A. Bolton, ed., *The Loma Prieta, California, Earthquake of October 17, 1989—Public Response.* U.S. Geological Survey Professional Paper 1553-B. Washington, D.C.: U.S. Government Printing Office.

27. Before 1990, earthquake insurance in California was available from private insurance companies for premiums on the order of $2 to $4 per thousand dollars of coverage, subject to a large deductible of 10 percent of insured value. Thus homeowners had to pay a premium of several hundred dollars a year and were still not covered against moderate damage within the 10 percent deductible (which, however, may be reduced by a federal income tax casualty loss deduction). Fewer than half the state's residents carried earthquake insurance. In 1990, after Loma Prieta, the state legislature established a mandatory program of minimum earthquake coverage as discussed in the text (Palm and Hodgson (1992), note 3, Ch. 2).

28. Even in Santa Clara County whose western boundary virtually coincides with the segment of the San Andreas Fault that ruptured in Loma Prieta, the prevalence of earthquake insurance reported by the survey respondents was only 51 percent in 1990. Only 29 percent of respondents in Contra Costa County, which experienced widespread effects of Loma Prieta, carried earthquake insurance, according to the follow-up survey. On the other hand, most victims would not have benefited greatly from such insurance since only 3.8 percent of residence damage reported for Santa Cruz County (adjoining Santa Clara County) was rated as "severe." "Slight" and "moderate" damage would presumably fall largely in the "10 percent deductible" and thus be borne by the owner. See P.W. O'Brien and D.S. Mileti (1993).

29. Palm and Hodgson, note 3.

30. R. Bolin and L.M. Stanford (1993). "Emergency Sheltering and Housing of Earthquake Victims: The Case of Santa Cruz County," in P. A. Bolton, ed. *The Loma Prieta, California, Earthquake of October 17, 1989—Public Response.* U.S. Geological Survey Professional Paper 1553-B. Washington, D.C.: U.S. Government Printing Office.

31. FEMA (1990), note 16, 27.

32. FEMA unpublished data.

33. Bolin and Stanford (1993), note 25, B 48.

34. FEMA (1990), note 16.

35. U.S. Geological Survey (1990). *Probabilities of Large Earthquakes in the San Francisco Bay Region, California.* USGS Circular 1053. Washington, D.C.: U. S. Government Printing Office.

36. Bay Area Regional Earthquake Preparedness Project, no date. *Living on the Fault.* (Public information brochure.) Oakland: BAREPP.

37. The epicentral region is the area of most intense ground motion on bedrock sites and was considered by the Working Group to extend for a radius of 10 kilometers (6 miles) from the earthquake fault rupture.

38. U.S. Geological Survey, no date. *The Next Big Earthquake in the Bay Area May Come Sooner Than You Think.* Menlo Park, Calif.: USGS.

39. Kenneth C. Topping (1992). *Oakland Hills Fire Prevention and Suppression Benefit Assessment District Report.* Pasadena, Calif.: Mimeo, 5.

40. Federal Emergency Management Agency (1992). *Hazard Mitigation Report for the East Bay Fire in the Oakland-Berkeley Hills.* Prepared pursuant to Disaster Declaration FEMA 919 DR CA. San Francisco: FEMA.

41. Edward J. Blakely (1992). "This City is Not for Burning," *Natural Hazards Observer* XVI(6): 1–3.

42. Blue Ribbon Fire Prevention Committee, for the East Bay Hill Urban-Wildland Interface Zone (1982). *Report.* Oakland: East Bay Regional Park District (mimeo), 64–68.

43. Ibid.

44. In view of the intensity of the 1991 Firestorm, it has been questioned whether the fuelbreak concept would have impeded the spread of the conflagration. The 1982 Blue Ribbon Committee perhaps overemphasized vegetation management to the exclusion of other mitigation approaches, such as coordination of fire fighting among Bay Area cities. (Personal communication, Tony Acosta, Oakland Parks Department, October 26, 1993.)

45. National Fire Protection Association, no date. *The Oakland/Berkeley Hills Fire— October 20, 1991.* Quincy, Mass.: NFPA.

46. California Governor's Office of Emergency Services, East Bay Hills Fire Operations Review Group (1992). *The East Bay Hills Fire: A Multi-Agency Review of the October 1991 Fire in the Oakland-Berkeley Hills.* Sacramento: OES.

47. The FEMA (1990) Loma Prieta report had urged the installation of on-site emergency generators by water supply managers, but the district had not done so by 1991. *The Oakland Tribune* (1991), 56.

48. National Fire Protection Association, note 40, 12.

49. Tony Acosta, Oakland Park District. Personal communication, October 26, 1993. News reports on the firestorms in Southern California in October 1993 mentioned a similar phenomenon of houses exploding from within.

50. Edward J. Blakely, Office of the Mayor, City of Oakland. Personal communication, August 12, 1993.

51. *East Bay Phoenix Journal,* October 18, 1993, 17.

52. Patricia Adler, ed. (1992). *Fire in the Hills: A Collective Remembrance.* Berkeley: Patricia Adler.

53. *The San Francisco Examiner,* October 31, 1993, E-1 and E-5.

54. Ibid.

55. *East Bay Phoenix Journal,* October 18, 1993, 16.

56. FEMA unpublished data.

57. FEMA(1992), note 40, 9.

58. The report reiterated the vegetation management recommendations of the 1982 Blue Ribbon Fire Prevention Committee whose executive summary was appended to the FEMA report.

59. California Governor's Office, note 46.

60. Subsequent research suggested that such measures may have been unnecessary and

possibly even counterproductive. See F.A. Booker, W.E. Dietrich, and L.M. Collins (1993). "Runoff and Erosion After the Oakland Firestorm." *California Geology* (Nov.-Dec.): 159–173.

61. The work group on this topic was chaired by Edward J. Blakely, Professor of City and Regional Planning at University of California, Berkeley, who served as policy advisor to the mayor of Oakland during the recovery period. He is now Dean of the School of Urban and Regional Planning at the University of Southern California.

62. Topping, note 35. The initiative in organizing regional response to vegetation management and fire response beyond the city of Oakland was assumed by the Hills Emergency Forum, an ongoing coalition of Oakland, Berkeley, East Bay Municipal Utility District, the East Bay Regional Park District, and the Lawrence Berkeley Laboratory of the University of California—all of which own and manage open lands in the Hills.

63. Rutherford H. Platt (1986). "Metropolitan Flood Loss Reduction Through Regional Special Districts," *Journal of the American Planning Association* (Autumn) 52(3): 467–479.

Conclusion and Recommendations

THE FEDERAL DISASTER LANDSCAPE: AN OVERVIEW

During the past 50 years, Congress has created a legal edifice of byzantine complexity to cope with natural disasters consisting of laws, agencies, programs, policies, and strategies, many of them intended to operate in "partnership" with state and local governments. Federal assistance is provided to households, businesses, farms, states, municipalities, special districts, and nongovernmental organizations under approximately 50 different laws and executive orders.[1]

The number and variety of federal agencies involved in disaster-related activities is breathtaking. For example, agencies interested in coastal hazards include FEMA, NOAA, the Army Corps of Engineers, the National Park Service (NPS), the Fish and Wildlife Service (FWS), and the U.S. Geological Survey (USGS). Issues of overlap, competition, and cross-purposes arise among different federal entities, adding to the cost and delays of accomplishing pre- and post-disaster actions. For instance, at Fire Island (Chapter 6), the Corps seeks to build an "interim" beach nourishment project based on assumptions disputed by recent research of USGS, and opposed by the FWS pending preparation of an environmental impact statement, with NPS caught in the middle trying to reconcile science and accommodation with the feisty Fire Islanders.

Financial transfers take many forms, particularly (1) direct federal grants and payments; (2) low-interest disaster loans; and (3) subsidized insurance payments. The cost to the federal treasury differs greatly among financial mechanisms and programs. Stafford Act grants involve a federal share of 75 percent except when the president raises the federal share to 90 or 100 percent for public assistance costs. States and local governments cover the nonfederal match requirements of federal grants, when necessary, in creative ways, sometimes even using other federal dollars to cover their share as in the Missouri buyout program. Low-interest disaster loans issued by the Small Business Administration are federally subsidized to the extent of the difference between the loan rate (about 4.0 percent as of June 1998) and the borrowing rate to the U.S. Treasury (currently about 5.6 percent). Loans not repaid have cost about two billion dollars since 1953. National Flood Insurance Program payments in average years are paid out of accrued premium revenue, but in large disasters additional funds are borrowed

277

from the Treasury. For FY 1986 through 1997, the NFIP experienced a net loss of $1.1 billion. With over $482 billion in total NFIP coverage as October 1, 1998—three-quarters of it in coastal counties—much larger program losses are probable in the event of major coastal disasters. Crop insurance payments are substantially subsidized by the government while agriculture disaster payments are 100 percent federal. Cost sharing requirements for other federal disaster-related programs vary widely.

In addition to financial transfers, federal agencies provide a number of direct services such as:

- Hurricane and tornado warnings by the National Weather Service
- Earthquake warnings by the U.S. Geological Survey (USGS)
- Mapping of hazard areas by FEMA, Tennessee Valley Authority (TVA), and USGS
- Communications, floodfighting, evacuation, and rescue activities by the Corps of Engineers, Coast Guard, and other military units
- Federal construction projects (e.g., Corps of Engineers beach nourishment activities)
- Watershed planning by the Natural Resources Conservation Service, Environmental Protection Agency, TVA, and other agencies
- Research and public information efforts supported by National Science Foundation, the Natural Oceanographic and Atmospheric Survey (NOAA), USGS, and other agencies

The Federal Response Plan completed by FEMA after Hurricane Andrew in 1992 provides a general framework for the disaster assignments of 26 federal agencies plus the American Red Cross. This plan attempts to coordinate the actions of hundreds of thousands of federal employees and billions of federal dollars that periodically are focused on natural disasters. But each of the major participating agencies is itself an organism of infinite and ever-changing complexity, comprising subagencies, offices, programs, divisions, branches, and units. Each major agency and its components operates within a unique web of executive, congressional, and judicial authority. Each responds to different sets of congressional committees, inquiries by members of Congress, and organizational lobbyists. Most are divided into headquarters and regional staff that often see their roles quite differently. Each undergoes frequent reorganization, engaging its staff in endless competition for budget, space, and personnel. And each federal agency possesses a distinctive culture, derived from its history, mission, leadership, headquarters location, and political connections.

FEMA, the lead federal disaster agency, by Washington standards is small, relatively new (established in 1979), with a modest regular annual appropriation (around $300 million). But FEMA administers billions of dollars of supplementary disaster funds, both directly through individual and public assistance grants and indirectly through mission assignments to other agencies, including the Corps of Engineers and the Departments of Transportation, Agriculture, Labor, Housing and Urban Development, and Commerce. FEMA is thus the broker of disaster resources to the much larger line agencies of the federal government. FEMA's political position has also been enhanced through the strong ties between Director James Lee Witt and President Clinton.

The cornucopia of federal disaster assistance is not without hidden costs. Seldom are the administrative or "overhead" costs of delivering federal services and benefits explicitly identified or estimated. Disasters in fact represent opportunities for many federal agencies. They provide unexpected funds through supplementary appropriations; they facilitate the hiring of additional staff and contractors; they provide opportunities to earn per diem and other benefits for time spent in the field (which may also be a personal hardship); they offer a respite from office politics; they may generate media publicity; and they offer a sense of doing something worthwhile to help people in need. Cynical or not, disasters are often good for the business of government as well as the reelection prospects of politicians. Even academic researchers such as the writer benefit by way of government grants to study disaster response and recovery. During the 1990s, the "business" of federal response to disasters has been brisk indeed.

But the importance of disaster programs to politicians and the federal bureaucracy recalls past periods of federal excess on behalf of worthy goals. For instance, the Cold War, western water development, flood control, interstate highways, urban housing, and even environmental programs have spawned vast agency fiefdoms and claims on the public treasury. According to James Wright, the former head of the TVA's Office of Floodplain Management Services: "The flood control construction program of the 1930–1950 era now seems to have been replaced by an equally massive federal relief and recovery assistance program for flood disasters in the present age."[2] Has disaster assistance become a new "pork barrel" to provide visible federal spending in localities and congressional districts that formerly were bestowed with dams, harbor improvements, sea walls, highways, military bases, and research labs? Even the hazard mitigation grant program (heretical though it is to suggest it) has the potential to become another billion dollar source of "nouveau pork."

Some related concerns: Is the present "system" the best way to deliver disaster assistance? Is it cost-effective and equitable? Is it efficient? Most important, for present purposes, how does it affect vulnerability? Does it undermine FEMA's goals to reduce vulnerability to natural hazards through mitigation? Does the high probability that any natural event—extreme or not—will be tacitly federally declared encourage building and rebuilding in hazardous areas?

These questions are deliberately disconcerting. Nothing except national security commands more public support than helping people and communities struck by disaster. It is a fundamental function of government—federal, state, and local—to protect, to rescue, to salvage, and to restore. Who can view images of the devastation of hurricanes, tornadoes, wildfires, or earthquakes and callously question the need for governmental help to the victims?

But natural disasters differ as to foreseeability. Wind disasters such as tornadoes and hurricanes like Andrew in 1992 strike randomly and viciously: Victims can do little to protect themselves from loss other than to take shelter if warnings sound. Seismic disasters are also capricious and diffuse in their effects. But building codes in areas of potential earthquake hazard have been proven to reduce damage to structures.[3] Urban-wildland fires as at Oakland in 1991 are fairly predictable, although regional fire hazards differ over time according to climate perturbations, as the 1998 Florida wildfires demonstrated. Most foreseeable are coastal disasters—flood, erosion, and wind—as well as riverine floods. The federal government has devoted over one billion dollars to mapping of flood hazard areas (albeit with varying degrees of accuracy) to inform communities and investors of the level of flood risk. Nevertheless, 40 percent ($2.58 billion) of all NFIP payments have pertained to 200,000 structures that have experienced repetitive losses: two or more claims while insured.[4] The message is not being received clearly that such properties are unsafe, or that the government seeks to discourage investment in such locations.

While some types of natural disasters thus strike randomly, others are geographically specific. Is it callous to call for potential victims—both individuals and communities—to exercise reasonable prudence to protect themselves from *foreseeable* and especially *repetitious* harm, even at the cost of foregoing optimal real estate profit and tax base enhancement for sites of known risk? If communities and private investors fail to act sensibly to protect themselves from natural hazards, to what extent should they be held harmless from the effects of their "own free choice?"

The issue is not whether federal assistance is fundamentally inappropriate, but rather when should it be provided, of what type and magnitude, and at whose cost? And as FEMA has rightly emphasized, how can recovery be designed to reduce further losses in the future through mitigation?

The remainder of this Conclusion will respond to these questions at two levels: micro and macro. At the micro level, the following section will review the lessons of the case studies of federal disaster assistance in three widely different hazards and locales. That will be followed by a discussion of broad-gauge policy options for federal disaster assistance in the new century.

LESSONS OF THE CASE STUDIES

The case studies on federal response to disasters at Fire Island, New York; St. Charles County, Missouri; and Oakland, California (Chapters 6, 7, and 8) disclose a double standard that afflicts national policy regarding natural hazards, particularly in areas of high amenity value such as coasts and steep hillsides. On the one hand, the federal government is increasingly expected to bear the major proportion of local and private economic costs of disasters through its grant, loan, and insurance programs. But on the other hand, government at all levels is stymied in preventing building and rebuilding in areas of known risk due to the property rights movement and fear of the "takings issue." According to a recent study of land use planning for natural hazards, "Unsustainable development practices (and large losses in natural disasters) have been the rule rather than the exception in urban areas across America."[5] Unwise building practices are often encouraged by federal development incentives, including even disaster assistance itself. In place of requiring potential victims and communities to mitigate their own future losses, at some immediate cost to themselves, mitigation is now becoming yet another item whose cost is expected to be assumed by the federal government, with no strings attached.

Fire Island

After the winter storms of 1992/1993 ravaged the oceanfront of Fire Island, New York, FEMA led an interagency hazard mitigation team whose report called for remapping and updating coastal hazard areas on the island. FEMA, however, has not followed its own advice. Fire Island as of June 1998 was still using flood insurance rate maps that predate the 1992/1993 storms. Similarly ignored was the team's recommendation that the state impose a moratorium on rebuilding within 25 feet landward of a bluff crest, back dune, or toe or vegetation line. The National Park Service has not updated its Dune District boundaries since 1976, with the result that its regulatory jurisdiction over the oceanfront lies out at sea. New York State has still not established erosion hazard area boundaries for Fire Island under its 1982 Coastal Erosion Hazard Area Act. Due in part to these and other regulatory failures, many oceanfront homes that were damaged in 1992/1993, or since, have been rebuilt on a more lavish scale. This process has been assisted by locally funded beach nourishment projects, which, in the absence of recent winter storms or hurricanes, have provided an illusion of stability over the last few years. But the realization that these projects are merely stopgaps inspires ongoing demands by the island for a federal $60 million "interim" beach nourishment project. If this project is built, additional oceanfront lots, currently on the open beach, would become eligible for reconstruction under current state and local laws. When storms again erode the nourished shoreline, those structures will be directly in the path of destruction. Even if the structures survive, the loss of sand beneath them will leave a row of opulent derelict homes standing in the surf zone, leading to new appeals to Congress for further beach nourishment. And so the cycle continues. . . .

grants (including hazard mitigation grants) are factored in, the total cost to the government may exceed the face value of the grant—certainly so when the non-federal share is reduced to 10 percent or zero.

Congress recognized this in creating the SBA disaster loan program in 1953, the National Flood Insurance Program in 1968, and updating the Federal Crop Insurance Program in 1980. While the federal subsidy for crop insurance has remained above 50 percent, the NFIP has gradually reduced its federal subsidy to near zero for an "average loss year" through rate increases and mitigation against lower-level floods. (Expected catastrophic NFIP coastal losses, however, could draw heavily on the federal treasury.) The SBA disaster loan program represents a business-like approach to disaster assistance for creditworthy victims who are uninsured or underinsured. Its average annual cost to the federal treasury is about $15 million in interest rate subsidies (1.5 percent of $1 billion in loans annually) plus an "actual loss rate" of 8.27 percent, amounting to a further federal cost of about $2 billion since 1953.

By comparison, the cost of direct federal disaster payments in the 1993 Midwest Flood alone—excluding federal insurance and loans—amounted to $4.1 billion. (See Table 7-5.)

Obviously, a shift of disaster costs from grants to insurance or loans would represent a significant gain for the federal treasury. It also would serve to enhance the sense of personal responsibility that federal "handouts" diminish. But participation rates in the crop insurance program is about 35 percent while NFIP policies cover only about 20 to 25 percent of the estimated structures at risk. However, NFIP coverage grew by 14 percent in fiscal 1998 to $482 billion, and the number of policies grew from 3.8 to 4.1 million, through an aggressive FEMA marketing effort and new sanctions under the 1994 Flood Insurance Reform Act.

A danger with any insurance program is adverse selection—namely, that those who purchase coverage are those most likely to incur a loss. Nearly 59 percent of NFIP policies and 63 percent of coverage are in coastal communities, particularly along the southeast Atlantic and Gulf of Mexico coasts, which have had heavy development during a period of low hurricane activity. An increase in hurricanes will yield much higher coastal NFIP losses in coming years. While coastal claims have accounted for fewer than half the losses paid by the program so far, oceanfront (V-zone) claims paid have averaged 2.5 times the level for other coastal or inland claims (Table 1-7).

Another difficulty with counting on the NFIP to offset rising Stafford Act outlays is that most of the latter are devoted to public assistance for infrastructure repair, which is not normally covered by flood insurance. Nor would flood insurance lower the cost of individual assistance, since persons eligible for IA would be less likely to own property covered by the NFIP. As matters stand, the NFIP and the Stafford Act apply to complementary sets of losses—an increase in NFIP coverage does not ipso facto represent a decrease in Stafford Act outlays. This is obvious when nonflood disasters are considered.

In the case of floods and hurricanes, Congress should amend the National Flood Insurance Act to cover public infrastructure. Just as communities carry insurance against civil liability or unexpected costs of snow removal, they should be encouraged and expected to insure against losses to their streets, parks, public buildings, and sewer and water systems from floods. As with earthquake insurance, such coverage might involve a sizeable deductible to keep the cost affordable. Communities would have to cover the deductible amount out of their operating budgets or "rainy day funds," if any.

Public assistance to communities under the Stafford Act should be reserved for (1) true catastrophes (e.g., Florida wildfires or Northridge Earthquake) where the capabilities of state and local governments are obviously inadequate, or (2) lower-income communities and counties suffering serious losses in less catastrophic events. In other words, public assistance should be need-based, like individual assistance. Communities capable of providing for their own reasonably foreseeable disaster losses should be required to do so; federal assistance would *supplement* local and state resources when absolutely necessary, as Congress has always specified in its disaster assistance laws.

A further measure to help tide over communities hit by unexpected short-term costs due to disasters would be a local government equivalent to the SBA disaster loan program. Communities should be able to borrow against future tax revenues at a below market interest rate. The same benefit should be extended to school districts and other special districts to assist them in repairing infrastructure damaged by a disaster, whether or not it is covered by a presidential declaration.

Reduce Federal Incentives to Develop or Redevelop in Hazardous Areas

The position of the federal government with respect to development in hazardous areas, especially along the coast, is ambivalent. On the one hand, the federal government is the prime source of disaster assistance and the principal champion of hazard mitigation to reduce future losses. On the other hand, the federal government fuels the momentum of building and rebuilding in hazardous areas as the unintended consequence of a variety of incentives, subsidies, and ad hoc projects. As stated by Peter May and Robert Deyle: "The net effect of federal programs is to encourage development in hazardous areas."[15] It thus undermines its own efforts to promote mitigation, rather like driving with the brakes on. As noted by Raymond Burby in remarks presented at the 1998 National Hurricane Conference:

The enormous losses experienced in hurricanes and the exposure to even more enormous losses in the future did not occur by accident. They are the result of conscious policy choices. The federal government has made it foolish (in fact, to seem like a sucker) for local government and individuals to take steps to avoid catastrophic losses in hurricanes. It has done this by violating each of five principles for reducing losses. Specifically, it has subsidized people to develop and occupy hazardous areas. It has provided only partial information about the risk of loss from hurricanes. It has done little to foster local commitment to manage land use to reduce risk. Its programs, which are confused and conflicting, resemble a house of cards built over time with little forethought or planning. And it has consistently avoided support of local land use planning and land management as hazard mitigation tools.[16]

Congress in the Coastal Barrier Resources Act,[17] as discussed in Chapter 3, acknowledged that federal benefits sometimes promote unwise development. Within coastal barriers designated by the act, it prohibited new flood insurance coverage and federally related financial assistance for the purchase or construction of buildings within CBRA units. It also banned federal grants or loans for the construction of new sewer and water supply systems, highways, airports, bridges, causeways, and shore stabilization projects. By 1998, the total extent of coastal areas subject to these restrictions amounted to nearly 1.3 million acres.[18]

But other coastal areas, already developed or developing, remain eligible for the full range of federal development incentives. Indeed pressure for growth in already developed areas may have been intensified by the nonavailability of federal benefits in designated CBRA units. Flood insurance under the NFIP in particular is viewed by property owners as a vested right of critical importance to the value of oceanfront structures, as reflected in the fight by Fire Island against proposed limits on NFIP coverage in areas of imminent erosion hazard. And federal tax and other incentives encourage building and rebuilding in other hazard areas across the nation, except where federally funded buyout programs have been implemented as in St. Charles County, Missouri.

Perhaps the most insidious federal stimulus to development along hazardous coasts is the federal tax code. Oceanfront real estate development is largely owned for investment purposes these days (rental and resale). As such, it often qualifies for federal (and perhaps state) tax deductions for expenses relating to investment property, including virtually all the costs of ownership—mortgage interest, taxes, maintenance, advertising, management. When personal use by the owner is limited to tax code guidelines or is nonexistent, annual depreciation can be deducted from the owner's taxable income. When a disaster declared by the president damages the property, a portion of the uninsured casualty loss may be deducted from taxable income as well. Detailed statistical evaluation of the effect of tax laws on development in hazardous areas is impossible due to the legal privilege of privacy accorded to individual tax returns, and the lack of any federal effort to aggregate the costs of specific classes of tax deductions.

Disaster assistance itself is an incentive to rebuilding in hazardous locations. Although direct federal assistance to homeowners is largely in the form of NFIP payments, the generous provision of public assistance to repair local infrastructure indirectly benefits property owners by avoiding new burdens on property taxes. After Topsail Island, North Carolina, was devastated by hurricanes Bertha and Fran in the summer of 1996, none of the beachfront communities that share the island had to raise taxes to pay for repair costs. Federal outlays averaged over $10,000 per full-time resident, according to one estimate.[19]

To shift community repair costs from disaster grants to insurance and loans, as proposed in the preceding section, would help to remove some of the unintended inducement of disaster assistance to rebuilding in areas of extreme hazard, such as coastal V-zones and riverine floodways. But federal flood insurance remains an uncertain factor in the rebuilding of coasts. To be sure, the NFIP stimulates elevation and more durable building practices when a substantially damaged structure is rebuilt. This raises the issue presented in the Midwest case study, namely how should substantial damage be measured, and by whom? Often the determination of substantial damage is made by local building inspectors who, especially in small beachfront communities, may be reluctant to force property owners to comply with current elevation requirements and thus hesitate to declare a structure "substantially damaged."

But apart from that issue, many structures designated as substantially damaged nevertheless experience further repetitive losses despite the NFIP standards for rebuilding. A recent study by the National Wildlife Federation (NWF) has found that repetitive claims (9568) on 10,921 substantially damaged properties *after* they were repaired actually exceeded slightly the number of claims (9523) *before* the substantial damage occurred. Total payments on properties subsequent to repair from a substantial damage incident amounted to $167 million as compared with $101 million in claims on the same properties prior to their being substantially damaged (not adjusted for inflation).[20] These findings cast doubt on the effectiveness of the NFIP substantial damage rule to protect structures from further losses.

Repetitive losses, whether or not involving substantially damaged structures, are a tremendous drain on the National Flood Insurance Program. According to the NFW study of national experience, repetitive claims (two or more) have been paid on only 2 percent of properties covered by flood insurance, but they accounted for 25 percent (200,182) of NFIP losses and comprised 40 percent ($2.58 billion) of NFIP payments. Those with three or more losses accounted for less than 1 percent of the policy base but made up 21.5 percent ($1.4 billion) of all flood insurance payments.[21]

These findings suggest that property owners can expect repetitive payments from the NFIP at little or no increase in premium cost or other penalty. Considering that automobile insurance rates usually rise after claims are paid or driving violations, it is remarkable that the National Flood Insurance Program never looks backward at the past claim record of a structure when setting premiums. This practice serves as yet another incentive to rebuild *in situ* and, in the absence of a substantial damage declaration, with essentially the same or greater degree of risk.

It is time to restrict the continued availability of modest-cost NFIP coverage in areas of extreme hazard from hurricanes, floods, and erosion. Existing coverage should remain in effect, but new coverage or renewal of repetitive-l oss property coverage should either be prohibited or a significantly higher premium should be charged. Current, the average annual premium for V-zones is $720, about twice the rate for the program as a whole. (See Table 1-7.) This is a bargain compared with the cost of general homeowner insurance, property taxes, condominium fees, and other costs of owning oceanfront property. It is also cheap in comparison with rents of $2000+ per week charged for many oceanfront properties during the vacation season. In other words, the cost of flood insurance, especially after a loss not involving substantial damage, is a stimulus to restoring the status quo ante. Losses experienced exceeding the maximum amount payable by the NFIP may be charged off against federal income taxes if damage occurs in a presidentially declared disaster. As remarked by a North Carolina state coastal manager, no other investments yield such high returns with so little financial risk as oceanfront rental property.[22]

Intensify Hazard Mitigation Through Land Use Controls and Incentives

A theme running through this book is the double bind of national disaster policy—the federal government is expected to come to the financial rescue of communities and individuals when disaster strikes, but is constrained by political pressure from requiring the latter to restrain unwise building or rebuilding in areas subject to recurrent natural hazards. The property rights movement and the takings issue discussed in Chapters 4 and 5 have diminished the ability or will of federal, state, and local officials to utilize land use regulations. National studies of risk reduction by several scholars document "noteworthy gaps in local policy adoption of land use measures for reducing risks posed by earthquakes, floods, and coastal hazards."[23] While the Supreme Court decision in *Lucas v. South Carolina Coastal Council* has not spawned an outpouring of anti-regulatory decisions in the lower federal or state courts, the threat of *Lucas*-based lawsuits has caused public agencies at all levels of government to refrain from restricting property rights, even in the face of well-documented natural hazards.[24]

This represents a retreat from land use regulation as a fundamental tool of hazard mitigation and prerequisite of federal assistance as provided in the National Flood Insurance Act of 1968,[25] the Flood Disaster Protection Act of 1973,[26] and the federal Disaster Assistance Act of 1974.[27] The declining prominence of land use regulation in hazard mitigation is reflected in its changing status in successive versions of the Unified National Program for Floodplain Management and in the National Mitigation Strategy of 1995. (See Chapter 3.)

FEMA has encouraged stronger building codes, particularly in areas subject to seismic and hurricane hazards. Structures along the coast that are substantially damaged by disasters are rebuilt higher and stronger than before. Building code regulations have so far escaped the wrath of the post-*Lucas* property rights movement.

But in the coastal context, building higher and stronger without horizontal setbacks ignores the effects of erosion. (See Chapter 6.) The NFIP has fostered a generation of costly private coastal structures, elevated on heavy pilings or columns, and eligible for flood insurance even though sometimes located literally at the mean high water line as permitted by FEMA V-zone regulations. Many of these structures are or soon will be vulnerable to undermining by rapid erosion and wave action, as during hurricanes Bertha and Fran, which struck North Carolina's coast in 1996. Unless government engages in costly, ongoing beach nourishment and shore protection, many of these structures will either collapse or become stranded in the surf zone as shorelines retreat. The resulting damage will impose costs on the NFIP, on private insurers for nonflood damage, and on federal taxpayers through casualty loss deductions and outlays for removal of debris and other assistance to individual owners and coastal communities. If a presidential disaster declaration is not involved, the local community could bear the cost of removing the debris from its shoreline. Unquantified but significant environmental and social costs may also arise from the cycle of building, damage, rebuilding, and final abandonment of coastal structures.

The tendency for coastal structures to be rebuilt more lavishly with government support has been frequently noted by news media, such as the following in *The New Yorker:* "The NFIP has, in effect, provided a financial safety net for developers to expand construction in [coastal] hazardous areas."[28]

FEMA responds that it is a "myth" that "the NFIP encourages coastal development."

> One of the NFIP's primary objectives is to guide development away from high flood-risk areas. NFIP regulations minimize the

impact of structures that are built in Special Flood Hazard Areas
(SFHAs) by requiring them not to cause obstructions to the natural flow of floodwaters. Also as a condition of community participation in the NFIP, those structures built within SFHAs must
adhere to strict floodplain management. . . . [29]

FEMA's faith in "strict floodplain management" by local governments to
"guide development away from high flood-risk areas" is open to question. Local
governments and states are not required to be any more strict than NFIP regulations require them to be. While some are relatively strict, many others are
not.[30] As related in Chapter 6, the goal of "strict floodplain management" at Fire
Island was compromised by FEMA's failure to follow the advice of its own hazard mitigation task force to redress flood insurance rate maps, including V-zones,
for the entire island.

Inadequacy of land use management as a tool of hazard mitigation is not limited to coastal hazard areas. The 1989 critique of national floodplain policy by
the ad hoc National Review Committee chaired by Gilbert F. White, quoted in
Chapter 3, bears repeating.

> The present status of floodplain management does not encourage
> complacency. The record is mixed. There are encouraging trends,
> as with the number of communities having some form of flood-
> plain regulations, but the rising toll of average annual flood losses
> has not been stopped or reversed. Some activities look more pro-
> ductive on paper than on the ground or in the real vulnerability
> of people. On balance, progress has been far short of what is desir-
> able or possible, or what was envisaged at times when the current
> policies and activities were initiated.[31]

The Association of State Floodplain Managers raised similar concerns in testimony to Congress in 1995.

> If we as a nation are going to bring escalating disaster costs under
> control we need to embrace the concept of hazard mitigation. . . .
> On an individual project basis, flood control projects have
> reduced flooding for design floods. But at the same time these
> policies have enticed additional development increasing the dam-
> age potential for severe floods, or have silently promoted the
> transfer of flood damages from one property to another. *Likewise,*
> *with a benevolent federal government, there has been little incentive at the*
> *local and state levels of government to minimize the creation of new flood*
> *hazards.*[32]

Finally, the 1995 Senate Bipartisan Task Force report also noted the ambiguity of the federal role: "Despite the federal government's apparently dominant functions, [it] has relatively little authority over critical components of loss control and emergency management such as . . . *the establishment and enforcement of land-use decisions and building codes.*"[33]

The experience of recovery from the Oakland Fire (Chapter 8) further underscores the weakness of land use planning and regulation in comparison with the pressure to rebuild with minor technical improvement of structural design.

These perceptions all point to the need to revitalize land use management in accordance with the original intent of Congress, as expressed in the major disaster assistance laws listed earlier. Much greater emphasis should be placed on restraining construction and reconstruction in coastal V-zones, in riverine floodways, and in areas of significant seismic or wildfire hazard. States and local governments should be expected to use their land use control powers to reduce hazard vulnerability. This should involve the full range of land use techniques, including buyouts (with or without federal assistance), subdivision regulations, wetlands restrictions, and floodplain zoning.[34]

The Community Rating System (CRS), initiated in 1989 and sanctioned by Congress in 1994, seeks to promote the use of such techniques, among other measures, by offering communities flood insurance premium reductions in exchange for actions exceeding minimum NFIP standards. So far, there is little evidence that CRS incentives have actually motivated communities to be stricter in their land use regulations. A drawback of the CRS approach is that it enshrines the existing NFIP regulations against revision, since communities now expect a premium reduction "bonus" for anything they are not currently required to do. Also CRS bonuses may be earned from a variety of less controversial actions than land use management, such as public education and warning systems.

Beyond incentives and cost-sharing of hazard mitigation projects, FEMA needs to demand that state and local recipients of federal assistance will enforce and where necessary strengthen their laws regulating development in hazardous areas. FEMA should work with the U.S. Department of Justice to seek a Supreme Court ruling that affirms the constitutionality of public land use regulations in areas of extreme hazard *regardless of the economic impact on the landowner and the community.* In other words, the dark cloud of the *Lucas* and *Dolan* decisions over hazard area restrictions, as discussed in Chapter 5, needs to be dispelled through judicial clarification.

The National Mitigation Strategy appropriately envisions a "partnership" of the federal, state, local, and private sectors in achieving reduction of natural hazard losses. This is consistent with the long-standing (but seldom honored in practice) principle that federal disaster relief is to *supplement* state and local capabilities. States and localities have legal authority to guide development away from the most hazardous locations and areas of repetitious damage. The federal government lacks this authority. It is therefore incumbent on the nonfederal members of the "partnership" to do their part in reducing future losses to their citizens, to their own jurisdictions, and to the nation. If they fail to act prudently in their development decisions, their eligibility for flood insurance and public assistance (PA) under the Stafford Act should be suspended, or provided under less favorable financial terms, such as at a higher premium level or nonfederal cost-share requirement.[35]

FEMA rightly advocates mitigation but, at the direction of Congress and the president, it lavishes federal disaster assistance benefits on victims regardless of their own willingness to protect themselves from loss and regardless of the wealth of recipient communities. The federal tax law with innumerable benefits for real estate investors and developers further offsets the economic pain of unwise land use decisions.

The message of this book was expressed in shorter form in a commentary by the writer on FEMA's National Mitigation Strategy, which views mitigation as the "cornerstone" of the nation's system of emergency management.

> FEMA's worthy goal to place mitigation at the heart of federal disaster policy is threatened by the larger political context of disaster response. On the one hand, the president and the Congress must be persuaded to exercise reasonable restraint in the federal assumption of disaster costs. On the other hand, and dependent upon that restraint, states and local governments must assume a meaningful share of the fiscal and political burden of mitigating hazards, including the willingness to use noncompensatory land use controls when necessary. If these adjustments in the larger context are not made, FEMA's mitigation efforts will not be a "cornerstone" but merely grains of sand scattered by the winds of political and private expedience.[36]

NOTES

1. Peter J. May and Robert E. Deyle (1998). "Governing Land Use in Hazardous Areas with a Patchwork System," in Raymond J. Burby, ed., *Cooperating with Nature: Confronting Natural Hazards with Land Use Planning for Sustainable Communities.* Washington, D.C.: Joseph Henry Press: 57–84, at 67.

2. James M. Wright (1996). "Effects of the Flood on National Policy: Some Achievements, Major Challenges Remain," in Stanley A. Changnon, ed., *The Great Flood of 1993: Causes, Impacts, and Responses.* Boulder, Colo.: Westview Press, 245–275, at 251.

3. David Alexander (1993). *Natural Disasters.* New York: Chapman and Hall, Ch. 5.

4. National Wildlife Federation (1998). *Higher Ground: A Report on Voluntary Property Buyouts in the Nation's Floodplains.* Washington, D.C.: NWF, 103.

5. Raymond J. Burby (1998). "Natural Hazards and Land Use: An Introduction," in

Raymond J. Burby, ed., *Cooperating with Nature: Confronting Natural Hazards with Land Use Planning for Sustainable Communities.* Washington, D.C.: Joseph Henry Press, 1–26, at 2.

6. Edward J. Kaiser and R. Matthew Goebel (1997). "Missouri after the Midwest Floods of 1993," in David R. Godschalk, ed., *Making Mitigation Work: Recasting Natural Hazards Planning and Implementation.* Final Report: National Science Foundation Grant No. CMS-9-9408322, Ch. 9. Chapel Hill: University of North Carolina Center for Urban and Regional Studies; Scott Faber (1996). *On Borrowed Land: Public Policies for Floodplains.* Cambridge, Mass.: Lincoln Institute of Land Policy, 10–11.

7. Faber, note 6, 11.

8. Thomas P. Zacharias (1996). "Impacts on Agricultural Production: Huge Financial Losses Lead to New Policies," in Stanley A. Changnon, ed., *The Great Flood of 1993: Causes, Impacts, and Responses.* Boulder, Colo.: Westview Press, 163–182, at 181.

9. The acquisition of such parcelized land is complicated but not unprecedented. The National Park Service acquired more than 4000 small lots in assembling land for the Indiana Dunes National Lakeshore in the 1970s.

10. See Rutherford H. Platt (1996). *Land Use and Society: Geography, Law, and Public Policy.* Washington, D.C.: Island Press, Ch. 9.

11. Richard A. Andrews, Director of California Office of Emergency Services, Remarks to National Research Council, Committee on Assessing the Costs of Natural Disasters, December 15, 1997.

12. An SBA declaration is based on at least 25 uninsured "substantial losses" to small businesses or homes within a particular city or county due to natural disaster, fire, or other mishap. Becky Brantley, SBA Loan Officer, pers. comm. June 19, 1998.

13. 42 USCA sec. 5176.

14. Edward J. Kaiser and R. Matthew Goebel (1997). "State Hazard Mitigation 409 Plans: Falling Short of Their Potential," in David R. Godschalk et al. (1997), *Making Mitigation Work: Recasting Natural Hazards Planning and Implementation.* Report to National Science Foundation on Grant No. CMS-9408322. Chapel Hill: University of North Carolina Center for Urban and Regional Studies.

15. May and Deyle, note 1, 67.

16. Raymond J. Burby (1998). "Programs that Sabotage Hurricane Mitigation Efforts (and What Can Be Done About Them)." Paper prepared for presentation to 1998 National Hurricane Conference, Norfolk, Va. (April 8–10).

17. 16 USCA secs. 1531–1543.

18. Coastal America (1998). *Coastal Challenges: A Guide to Coastal and Marine Issues.* Washington, D.C.: Coastal America, 125.

19. Craig Whitlock (1997). "Flooded with Generosity," *The News and Observer* (Raleigh, N.C.). November 9, 1997, 1.

20. National Wildlife Federation, note 4, Table 3. VIII.

21. Ibid., 103.

22. Anonymous personal communication, June 17, 1998.

23. May and Deyle, note 1, 60–61.

24. Peter Douglas, Executive Director, California Coastal Commission, pers. comm., April 20, 1998.

25. PL 90-448, Sec. 1361: "[FEMA] shall . . . develop comprehensive criteria to encourage . . . the adoption of adequate state and local measures which . . . will (1) con-

strict the development of land which is exposed to flood damage . . . (2) guide the development of proposed construction away from locations which are threatened by flood hazards . . ." (42 USCA, Sec. 4102).

26. PL 93-234, Sec. 2(b): "The purpose of this act . . . is to (3) require states or local communities, as a condition of future federal financial assistance, to participate in the flood insurance program and to adopt adequate flood plain ordinances with effective enforcement provisions consistent with federal standards to reduce or avoid future flood losses . . ." (42 USCA, Sec. 4002(b)(3)).

27. PL 93-288, Sec. 406: "As a condition of any disaster loan or grant made under . . . this act . . . the state or local government shall agree that the natural hazards in the areas in which the proceeds of the grant or loans are to be used shall be evaluated and appropriate action shall be taken to mitigate such hazards, including safe land use and construction practices, in accordance with [federal] standards. . . ." (Renumbered Sec. 409 in the Stafford Act, PL 100-707; 42 USCA Sec. 5176.)

28. Wade Graham, "Beachless." *The New Yorker* December 16, 1996, 66.

29. FEMA, "Myths and Facts about the National Flood Insurance Program." FEMA Website, June 1998.

30. James M. Holway and Raymond J. Burby (1993). "Reducing Flood Losses: Local Planning and Land Use Controls." *Journal of the American Planning Association* 59(2): 205–216.

31. National Review Committee (1989). "Action Agenda for Managing the Nation's Floodplains" (mimeo). Reprinted as Appendix F of *Floodplain Management in the United States*, Vol. II. Washington, D.C.: U.S. Government Printing Office.

32. Doug Plasencia. Testimony on behalf of the Association of State Floodplain Managers, Inc., before the U.S. Senate Committee on Environment and Public Works Subcommittee on Transportation and Infrastructure, February 14, 1995. (Emphasis added.)

33. Senate Bipartisan Task Force on Disaster Relief Funding (1995). *Federal Disaster Assistance.* Washington, D.C.: U.S. Government Printing Office.

34. Platt, note 10; Raymond Burby, ed. (1998). *Cooperating with Nature: Confronting Natural Hazards with Land Use Planning for Sustainable Communities.* Washington, D.C.: Joseph Henry Press.

35. As stated above, individual assistance (IA) should not be jeopardized by political considerations, including the unwillingness of a community to enforce stringent land use regulations in hazardous areas.

36. Rutherford H. Platt (1996). "Hazard Mitigation: Cornerstone or Grains of Sand?" *Natural Hazards Observer* (September): 10–11.

SELECTED BIBLIOGRAPHY
FOR FURTHER READING

BOOKS

Alexander, David. 1995. *Natural Disasters.* London: Chapman and Hall.

————. 1999. *Natural Disasters in the Modern World.* London: Terra Publishing.

Beatley, Timothy, David J. Brower, and Anna K. Schwab. 1994. *An Introduction to Coastal Zone Management.* Washington, D.C.: Island Press.

Berke, Philip R., and Timothy Beatley. 1997. *After the Hurricane: Linking Recovery to Sustainable Development in the Caribbean.* Baltimore: Johns Hopkins University Press.

Blaikie, Piers, Terry Cannon, Ian Davis, and Benjamin Wisner. 1994. *At Risk: Natural Hazards, People's Vulnerability, and Disasters.* London: Routledge.

Bronson, William. 1959/1989. *The Earth Shook, The Sky Burned.* San Francisco: Chronicle Books.

Burby, Raymond J., and Steven P. French. 1985. *Flood Plain Land Use Management: A National Assessment.* Boulder, Colo.: Westview Press.

Burby, Raymond J., S.A. Bollens, J.M. Holloway, E.J. Kaiser, D. Mullan, and J.R. Sheaffer. 1988. *Cities Under Water: A Comparative Evaluation of Ten Cities' Efforts to Manage Floodplain Land Use.* Program on Environment and Behavior Monograph No. 47. Boulder, Colo.: Institute of Behavioral Science of the University of Colorado.

Burby, Raymond J., Beverly Ciglier, Steven P. French, Edward Kaiser, Dale Roenigk, Dana West, and Dale Whittington. 1991. *Sharing Environmental Risks: How to Control Governments' Losses in Natural Disasters.* Boulder, Colo.: Westview Press.

Burby, Raymond J., ed. 1998. *Cooperating with Nature: Confronting Natural Hazards with Land Use Planning for Sustainable Communities.* Washington, D.C.: Joseph Henry/National Academy Press.

Burton, Ian, Robert W. Kates, and Gilbert F. White. 1993. *The Environment as Hazard* (2nd ed.). New York: Guilford Press.

Burton, Ian, Robert W. Kates, and Rodney E. Snead. 1969. *Human Ecology of Coastal Flood Hazard in Megalopolis.* Research Paper No. 115. Chicago: University of Chicago Department of Geography.

Bush, David M., Orrin H. Pilkey Jr., and William J. Neal. 1996. *Living by the Rules of the Sea.* Durham, N.C.: Duke University Press.

Changnon, Stanley A., ed. 1996. *The Great Flood of 1993: Causes, Impacts, and Responses.* Boulder, Colo.: Westview Press.

Comfort, Louise K., ed. 1988. *Managing Disaster: Strategies and Policy Perspectives.* Durham, N.C.: Duke University Press.

301

Cuny, Frederick C. 1983. *Disasters and Development.* New York: Oxford University Press.

Cutter, Susan, ed. 1994. *Environmental Risks and Hazards.* Englewood Cliffs, N.J.: Prentice-Hall.

Daniel, Pete. 1977. *Deep'N As It Come: The 1927 Mississippi River Flood.* New York: Oxford University Press.

Drabek, Thomas E. 1986. *Human Systems Responses to Disaster.* New York: Springer-Verlag.

Drabek, Thomas, and G. Hoetmer, eds. 1991. *Emergency Management: Principles and Practices for Local Government.* Washington, D.C.: International City Management Association.

Fischel, William. 1995. *Regulatory Takings: Law, Economics, and Politics.* Cambridge, Mass.: Harvard University Press.

Friesema, H. Paul, James Caporaso, Gerald Goldstein, Robert Lineberry, and Richard McCleary. 1979. *Aftermath: Communities After Natural Disasters.* Beverly Hills and London: Sage Publications.

Godschalk, David R., Timothy Beatley, Philip Berke, David Brower, Edward Kaiser, Charles Bohl, and R. Matthew Goebel. 1998. *Natural Hazard Mitigation: Recasting Disaster Policy and Planning.* Washington, D.C.: Island Press.

Haas, J. Eugene, Robert W. Kates, and Martyn J. Bowden, eds. 1977. *Reconstruction Following Disaster.* Cambridge, Mass.: The MIT Press.

Handmer, John, ed. 1987. *Flood Hazard Management: British and International Perspectives.* Norwich, England: Geo Books.

Hewitt, Kenneth, ed. 1983. *Interpretations of Calamity.* Boston: Allen and Unwin.

Hoyt, William G., and Walter B. Langbein. 1955. *Floods.* Princeton, N.J.: Princeton University Press.

Kates, Robert W., and Ian Burton, eds. 1986. *Geography, Resources, and Environment: Themes from the Work of Gilbert F. White.* Chicago: University of Chicago Press, pp. 28–68.

Kaufman, Wallace, and Orrin H. Pilkey, Jr. 1979. *The Beaches Are Moving: The Drowning of America's Shoreline.* Durham, N.C.: Duke University Press.

Kelley, Robert. 1989. *Battling the Inland Sea: American Political Culture, Public Policy, and the Sacramento Valley, 1850–1986.* Berkeley: University of California Press.

Kirby, Andrew, ed. 1990. *Nothing to Fear: Risks and Hazards in American Society.* Tucson: University of Arizona Press.

Kunreuther, Howard, R. Ginsberg, L. Miller, P. Slovic, B. Borkan, and N. Katz. 1978. *Disaster Insurance Protection: Public Policy Lessons.* New York: John Wiley.

Kunreuther, Howard, and Richard J. Roth, Jr., eds. 1998. *Paying the Price: The Status and Role of Insurance Against Natural Disasters in the United States.* Washington, D.C.: National Academy of Sciences Press.

Kusler, Jon. 1993. *Public Liability and Natural Hazards: The Common Law and Regulatory Takings.* Washington, D.C.: Omni Press for the National Science Foundation.

Leopold, Luna B., and Thomas Maddock. 1954. *The Flood Control Controversy: Big Dams, Little Dams, and Land Management.* New York: The Ronald Press Company.

May, Peter J. 1985. *Recovering From Catastrophes: Federal Disaster Relief Policy and Politics.* Westport, Conn.: Greenwood Press, Table 2.1.

May, Peter J., and Walter Williams. 1986. *Disaster Policy Implementation: Strategies Under Shared Governance.* New York: Plenum Press.

May, Peter J., Raymond J. Burby, J.E. Dixon, Neil J. Ericksen, John Handmer, S. Michaels, and D.I. Smith. 1996. *Environmental Management and Governance: Intergovernmental Approaches to Hazards and Sustainability.* London and New York: Routledge.

McCullough, David G. 1968. *The Johnstown Flood.* New York: Simon and Schuster.

Meltz, Robert, Dwight H. Merriam, and Richard M. Frank. 1999. *The Takings Issue: Constitutional Limits on Land Use Control and Environmental Regulation.* Washington, D.C.: Island Press.

Mileti, Dennis S. 1999. *Disasters by Design: A Reassessment of Natural Hazards in the United States.* Washington, D.C.: Joseph Henry Press.

Mitchell, James K. 1974. *Community Response to Coastal Erosion: Individual and Collective Adjustments to Hazard on the Atlantic Shore.* Research Paper No. 156. Chicago: University of Chicago Department of Geography.

Mitchell, James K., ed. 1999. *Crucibles of Hazard: Megacities and Disasters in Transition.* Tokyo: United Nations University Press.

Moore, Jamie W., and Dorothy P. Moore. 1989. *The Army Corps of Engineers and the Evolution of Federal Flood Plain Management Policy.* Program on Environment and Behavior Special Pub. No. 20. Boulder, Colo.: Institute of Behavioral Science.

National Research Council. 1990. *Managing Coastal Erosion.* Washington, D.C.: National Academy Press.

———. 1995. *Beach Nourishment and Protection.* Washington, D.C.: National Academy Press.

———. 1995. *Flood Risk Management and the American River Basin: An Evaluation.* Washington, D.C.: The National Academy Press.

Palm, Risa. 1990. *Natural Hazards: An Integrative Framework for Research and Planning.* Baltimore: Johns Hopkins University Press.

Parker, Dennis, and John Handrner, eds. 1992. *Hazard Management and Emergency Planning: Perspectives on Britain.* London: James and James.

Peacock, Walter, Betty Hearn Morrow, and Hugh Gladwin, eds. 1997. *Hurricane Andrew: Ethnicity, Gender, and the Sociology of Disaster.* London: Routledge.

Petak, William J., and Arthur A. Atkisson. 1982. *Natural Hazard Risk Assessment and Public Policy: Anticipating the Unexpected.* New York: Springer-Verlag.

Pielke, Roger A., Jr. 1996. *Midwest Flood of 1993: Weather, Climate, and Societal Impacts.* Boulder, Colo.: National Center for Atmospheric Research, Environmental and Societal Impacts Group.

Pielke, Roger A., Jr., and Roger A. Pielke, Sr. 1997. *Hurricanes: Their Nature and Impacts on Society.* London: John Wiley.

Pilkey, Orrin H., and Katharine L. Dixon. 1996. *The Corps and the Shore.* Washington, D.C.: Island Press.

Platt, Rutherford H. et al. 1980. *Intergovernmental Management of Floodplains.* Program and Technology, Environment, and Man Monograph No. 30. Boulder, Colo.: University of Colorado Institute of Behavioral Science.

Platt, Rutherford H., Sheila G. Pelczarksi, and Barbara K.R. Burbank, eds. 1987. *Cities on the Beach: Management Issues of Developed Coastal Barriers.* Research Paper No. 224. Chicago: University of Chicago Department of Geography Research Series.

Platt, Rutherford H., ed. 1987. *Regional Management of Metropolitan Floodplains.* Program and Technology, Environment and Man Monograph No. 45. Boulder, Colo.: University of Colorado Institute of Behavioral Science.

Platt, Rutherford H. et al. 1992. *Coastal Erosion: Has Retreat Sounded?* Program on Behavior and Environment Monograph No. 53. Boulder, Colo.: University of Colorado Institute of Behavioral Science.

Platt, Rutherford H. 1996. *Land Use and Society: Geography, Law, and Public Policy.* Washington, D.C.: Island Press.

Rossi, Peter, James D. Wright, Eleanor Weber-Burdin, and Joseph Pereira. 1983. *Victims of the Environment: Loss from Natural Hazards in the United States, 1970–1980.* New York and London: Plenum Press.

Rosen, Henry, and Martin Reuss, eds. 1988. *The Flood Control Challenge: Past, Present, and Future.* Chicago: Public Works Historical Society.

Rubin, Claire et al. 1985. *Community Recovery from a Major Natural Disaster.* Program on Environment and Behavior Monograph No. 41. Boulder, Colo.: University of Colorado Institute of Behavioral Science.

Quarantelli, E. L., ed. 1998. *What Is a Disaster? Perspectives on the Question.* London: Routledge.

Schwab, James, ed. In Press. *Planning for Post-Disaster Recovery and Reconstruction.* Planning Advisory Report. Chicago: American Planning Association.

Showalter, Pamela S., William E. Riebsame, and Mary Fran Myers. 1993. *Natural Hazard Trends in the United States: A Preliminary Review for the 1990s.* Natural Hazards Research and Applications Information Center. University of Colorado at Boulder.

Smith, Keith. 1996. *Environmental Hazards: Assessing Risk and Reducing Disaster.* New York: Routledge.

Sylves, Richard T., and William H. Waugh, Jr., eds. *Disaster Management in the U.S. and Canada.* Springfield, Ill.: Charles C. Thomas.

Thomas, Gordon, and Max Morgan Witt. 1971. *The San Francisco Earthquake.* New York: Stein and Day.

Tobin, Graham A., and Burrell E. Montz. 1994. *The Great Midwestern Floods of 1993.* Philadelphia: Saunders College Publishing.

Tobin, Graham A., and Burrell E. Montz. 1997. *Natural Hazards: Explanation and Integration.* New York: Guilford Press.

Turner, B.L., II, William C. Clark, Robert W. Kates, John F. Richards, Jessica T. Mathews, and William B. Meyer, eds. 1990. *The Earth as Transformed by Human Action: Global and Regional Changes in the Biosphere over the Past 300 Years.* New York: Cambridge University Press.

White, Gilbert F. 1945. *Human Adjustment to Floods: A Geographical Approach to the Flood Problem in the United States.* Research Paper No. 29. Chicago: University of Chicago Department of Geography.

———. 1975. *Flood Hazard in the United States: A Research Assessment.* NSF-RA-E-75-006. Boulder, Colo.: University of Colorado Institute of Behavioral Science.

White, Gilbert F., ed. 1964. *Choice of Adjustment to Floods.* Research Paper No. 93. Chicago: University of Chicago Department of Geography.

————. ed. 1974. *Natural Hazards: Local, National, and Global.* New York: Oxford University Press.

Yanev, Peter I. 1990. *Peace of Mind in Earthquake Country.* San Francisco: Chronicle Books.

BOOK CHAPTERS, SYMPOSIA, AND ARTICLES

Allen, W.H., 1993. "The Great Flood of 1993." *BioScience* 43(11): 732.

Association of State Floodplain Managers. 1983. *Preventing Coastal Flood Disasters: The Role of the States and Federal Response.* Special Pub. No. 7. Boulder, Colo.: Natural Hazards Research and Applications Information Center.

————. 1984. *Managing High Risk Flood Areas: 1985 and Beyond.* Special Pub. No. 11. Boulder, Colo.: Natural Hazards Research and Applications Information Center.

————. 1997. *Floodplain Management in a Multifaceted World.* Boulder, Colo.: Natural Hazards Research and Applications Information Center.

Baumann, Duane. 1994. Symposium: "Coping with the Flood: the Next Phase." *Water Resources Update,* Issue No. 94. (entire issue).

Berke, Philip R., Jack Kartez, and Dennis Wenger. 1993. "Recovery After Disaster: Achieving Sustainable Development, Mitigation, and Equity." *Disasters* 17(2): 93–109.

Berke, Philip R. 1998. "Reducing Natural Hazard Risk Through State Growth Management." *Journal of the American Planning Association* 64(1): 76–87.

Berke, Philip R., Dale Roenigk, Edward Kaiser, and Raymond J. Burby. 1996. "Enhancing Plan Quality: Evaluating the Role of State Planning Mandates for Natural Hazard Mitigation." *Journal of Environmental Planning and Management* 39(1): 79–96.

Bolin, Robert and L. Stanford. 1991. "Shelter, Housing, and Recovery: A Comparison of U.S. Disasters." *Disasters* 45(1): 25–34.

Boliens, Scott A., Edward J. Kaiser, and Raymond J. Burby. 1988. "Evaluating the Effects of Local Floodplain Management Policies in Property Owner Behavior." *Environmental Management* 12(3): 311–325.

————. 1989. "Private Profit and Public Safety as Outcomes of Local Land Use Regulation." *Environment and Planning B: Planning and Design* 16(1): 7–22.

Boswell, Michael R., Robert D. Deyle, Richard A. Smith, and E. Jay Baker. 1999. "A Quantitative Method for Estimating Probable Public Costs of Hurricanes." *Environmental Management* 23(3). In press.

Clary, Bruce B. 1985. "The Evolution and Structure of Natural Hazard Policies." *Public Administration Review* (45): 20–28.

Deyle, Robert E., Steven P. French, Robert B. Olshansky, and Robert G. Paterson. 1998. "Hazard Assessment: A Factual Basis for Planning and Mitigation." In Raymond J. Burby, ed., *Cooperating with Nature: Confronting Natural Hazards with Land-Use Planning for Sustainable Communities.* Washington, D.C.: Joseph Henry Press, pp. 119–166.

Dunham, Allison. 1959. "Flood Control Via the Police Power." *University of Pennsylvania Law Review* 107(4): 1098–1131.

Galloway, Gerald E. 1997. "River Basin Management in the 21st Century: Blending Development with Economic, Ecologic, and Cultural Sustainability." *Water International* (22): 82–89.

Godschalk, David R., and Kathryn Cousins, eds. 1985. Symposium: "Coastal Management: Planning on the Edge." *Journal of the American Planning Association* 51(3): 263–336.

Holway, James M., and Raymond J. Burby. 1993. "Reducing Flood Losses: Local Planning and Land Use Controls." *Journal of the American Planning Association* 59(2): 205–216.

Kunreuther, Howard, and L. Miller. 1985. "Insurance Versus Disaster Relief: An Analysis of Interactive Modelling for Disaster Policy Planning." *Public Administration Review* 45(2): 147–154.

Kusler, Jon A., and Larry Larson. 1993. "Beyond the Ark: A New Approach to U.S. Floodplain Management." *Environment* 35(5): 6–11; 31–34.

May, Peter J., and Patricia Bolton. 1986. "Reassessing Earthquake Hazards Reduction Measures." *Journal of the American Planning Association* 52(4): 443–451.

May, Peter J., and Thomas A. Birkland. 1994. "Earthquake Risk Reduction: An Examination of Local Regulatory Efforts." *Environmental Management* 18(6): 923–937.

May, Peter J., 1997. "State Regulatory Roles: Choices in the Regulation of Building Safety." *State and Local Government Review* 29(1): 70–80.

May, Peter J., and Robert E. Deyle. 1998. "Governing Land Use in Hazardous Areas with a Patchwork System." In Raymond J. Burby (ed.), *Cooperating with Nature: Confronting Natural Hazards with Land-Use Planning for Sustainable Communities.* Washington, D.C.: Joseph Henry Press, pp. 57–84.

May, Peter J., Raymond J. Burby, and Howard Kunreuther. 1998. "Policy Design for Earthquake Hazard Mitigation: Lessons from Energy Conservation, Radon Reduction, and Termite Control." *Earthquake Spectra-Journal of the International Association Earthquake Engineering Research Institute* (In press.)

Montz, Burrell E., and Eve C. Gruntfest. 1986. "Changes in American Urban Floodplain Occupance since 1958: The Experiences of Nine Cities." *Applied Geography* 6: 325–338.

Myers, Mary Fran, and Gilbert F. White. 1993. "The Challenge of the Mississippi Flood." *Environment* 35(10): 6–9; 25–36.

Platt, Rutherford H. 1976. "The National Flood Insurance Program: Some Midstream Perspectives." *The Journal of the American Institute of Planners* 42(2): 303–313.

————. 1978. "Coastal Hazards and National Policy: A Jury-Rig Approach." *Journal of the American Institute of Planners* 42(1): 170–180.

————. 1982. "The Jackson Flood of 1979: A Public Policy Disaster." *Journal of the American Planning Association* 48(2): 219–231.

————. 1986. "Metropolitan Flood Loss Reduction Through Regional Special Districts." *Journal of the American Planning Association* 52(4): 467–479.

————. 1994. "The Evolution of Coastal Hazards Policies in the United States." *Coastal Management* 22(3): 265–284.

————. 1995. "Report on Reports: Sharing the Challenge—Floodplain Management into the 21st Century." *Environment* 37(1): 25–29.

————. 1995. "The Mississippi River Basin—Crucible of National Flood Policies." In J. Gardner, 0. Starololszky, and V. Yevjevich, eds., *Defence from Floods and Floodplain Management.* Dordrecht, The Netherlands: Kluwer Academic Publishers, pp. 381–394.

Platt, Rutherford H., and S. Lawrence Dingman. 1977. "Floodplain Zoning: Implications of Hydrologic and Legal Uncertainty." *Water Resources Research* 13(3): 519–523.

Platt, Rutherford H., Timothy Beatley, and H. Crane Miller. 1991. "The Folly at Folly Beach and Other Failings of U.S. Coastal Erosion Policy." *Environment* 33(9): 6–9; 25–32.

Riley, Ann L. 1994. "The Greening of Federal Flood-Control Policies: The Wildcat–San Pablo Creeks Case." In Rutherford H. Platt, Rowan A. Rowntree, and Pamela C. Muick, eds., *The Ecological City: Preserving and Restoring Urban Biodiversity.* Amherst: University of Massachusetts Press, pp. 217–230.

Tobin, Graham A., and Burrell E. Montz. 1988. "Catastrophic Flooding and the Response of the Real Estate Market." *The Social Science Journal* 25(2): 167–177.

Strong, Ann Louise, Daniel R. Mandelker, and Eric Damian Kelly. 1996. "Property Rights and Takings." *Journal of the American Planning Association* 62(1): 5–16.

White, Gilbert F., ed. 1986. Symposium: "Defusing Natural Disasters." *Journal of the American Planning Association* 52(4): 429–488.

White, Gilbert F. 1997. "The River as a System: A Geographer's View of Promising Approaches." *Water International* 22: 79–81.

REPORTS AND GOVERNMENT DOCUMENTS

Faber, Scott. 1996. *On Borrowed Land: Public Policies for Floodplains.* Cambridge, Mass.: Lincoln Institute of Land Policy.

Federal Emergency Management Agency. 1995. *National Mitigation Strategy: Partnerships for Building Safer Communities.* Washington, D.C.: Federal Emergency Management Agency.

———. 1997. *Report on Costs and Benefits of Natural Hazard Mitigation.* Washington, D.C.: Federal Emergency Management Agency Mitigation Directorate.

Federal Interagency Floodplain Management Task Force. 1992. *Floodplain Management in the United States: An Assessment Report.* Vol. I - Summary Report. Boulder, Colo.: University of Colorado Natural Hazards Information Center, 1992. Vol. II - Full Report Washington, D.C.: U.S. Government Printing Office.

Godschalk, David R. 1998. *Coastal Hazards Mitigation: Public Notification, Expenditure Limitations, and Hazard Areas Acquisition.* Report prepared for the North Carolina Division of Coastal Management. Chapel Hill: University of North Carolina Center for Urban and Regional Studies.

Insurance Institute for Property Loss Reduction. 1995. *Coastal Exposure and Community Protection: Hurricane Andrew's Legacy.* Boston: IIPLR.

Interagency Floodplain Management Review Committee. 1994. *Sharing the Challenge: Floodplain Management into the 21st Century.* Washington, D.C.: U.S. Government Printing Office.

Library of Congress, Congressional Research Service. 1990. *When the United States Takes Property: Legal Principles.* Congressional Research Service Report for Congress. Washington, D.C.: Congressional Research Service.

———. 1992. *Federal Emergency Management Agency and the Disaster Relief Fund.* Congressional Research Service Report for Congress. Washington, D.C.: Congressional Research Service.

————. 1994. *A Descriptive Analysis of Federal Relief, Insurance, and Loss Reduction Programs for Natural Hazards.* Congressional Research Service Report for Congress. Washington, D.C.: Congressional Research Service.

National Academy of Public Administration. 1993. *Coping with Catastrophe.* Washington, D.C.: National Academy of Public Administration, National Performance Review.

National Performance Review. 1993. *Creating a Government that Works Better and Costs Less: Federal Emergency Management Agency.* Washington, D.C.: Office of the Vice President.

National Research Council. 1991. *A Safer Future: Reducing the Impacts of Natural Disasters.* Washington, D.C.: National Academy Press.

————. 1994. *Facing the Challenge: The U.S. National Report to the IDNDR Conference on Natural Disaster Reduction.* Washington, D.C.: National Academy Press.

National Review Committee. 1992. *Action Agenda for Managing the Nation's Floodplains.* Special Publication No. 25. Boulder, Colo.: University of Colorado Institute of Behavioral Science.

National Wildlife Federation. 1998. *Higher Ground.* Washington, D.C.: National Wildlife Federation.

Task Force on Federal Flood Control Policy. 1966. *A Unified National Program for Managing Flood Losses.* House Doc. 465 (89th Cong. 2nd sess.) Washington, D.C.: U.S. Government Printing Office.

U.S. General Accounting Office. 1980. *Federal Disaster Assistance: What Should the Policy Be?* PAD-80-39. Washington, D.C.: General Accounting Office.

————. 1982. *National Flood Insurance: Marginal Impact on Flood Plain Development— Administrative Improvements Needed.* GAO/RCED-82-105. Washington, D.C.: General Accounting Office.

————. 1991. *Disaster Assistance: Federal, State, and Local Responses to Natural Disasters Need Improvement.* GAO/RCED-91-43. Washington, D.C.: General Accounting Office.

————. 1993. *Disaster Management: Improving the Nation's Response to Catastrophic Disasters—Report to Congressional Requesters.* Report No. GAO/RCED-93-186, B-253822. Washington, D.C.: General Accounting Office.

————. 1994. *Flood Insurance: Financial Resources May Not Be Sufficient to Meet Future Expected Losses.* GAO/RCED-94-80. Washington, D.C.: General Accounting Office.

————. 1996. *Disaster Assistance: Improvements Needed in Determining Eligibility for Public Assistance.* GAO/RCED-96-113. Washington, D.C.: General Accounting Office.

U.S. Geological Survey (USGS). 1993. *The Loma Prieta, California, Earthquake of October 17, 1989—Public Response.* Patricia A. Bolton, ed., USGS. Professional Paper I 553-B. Washington, D.C.: U.S. Government Printing Office.

U.S. House of Representatives Bipartisan Natural Disasters Task Force. 1994. *Report.* (December 14). Mimeo.

U.S. Senate Bipartisan Task Force on Funding Disaster Relief. 1995. *Federal Disaster Assistance.* Washington, D.C.: U.S. Government Printing Office.

ABOUT THE CONTRIBUTORS

MIRIAM GRADIE ANDERSON is currently a regional planner with the Massachusetts Department of Environmental Management Flood Hazard Management Program. Between 1993 and 1997, she worked as an associate planner for St. Charles County, Missouri. Ms. Anderson received a master's in geography from the University of Massachusetts at Amherst in 1987.

ALEXANDRA D. DAWSON is an attorney specializing in environmental and land use law. She is a professor emerita at Antioch New England University and co-executive director of the Water Supply Citizens Advisory Committee. She serves as legal affairs director for the Massachusetts Association of Conservation Commissions. Among her publications are a book, *Land Use Planning and the Law,* and a chapter on "Wetland and Floodplain Law" in Rathkopf, *Law of Zoning and Planning* (West).

UTE J. DYMON is an associate professor of geography at Kent State University in Ohio. She received her Ph.D. in geography from Clark University in Worcester, Massachusetts. The application of geographical information systems in response to extreme hazardous events is currently the focus of her research and teaching. From her former studies of hurricanes Andrew and Fran, her most recent research concerns the spatial issues arising from the emergency evacuation of sensitive populations. She has published widely on the use of maps and information technology during and after major disasters.

JESSICA SPELKE JANSUJWICZ holds a B.A. in zoology from Connecticut College and an M.S. in wildlife and fisheries conservation from the University of Massachusetts at Amherst. Her research interests include the ecology of New England salt marshes and public policy issues related to the conservation and management of coastal resources, particularly wetlands. Jessica has spent the past year teaching environmental education at a science museum in Vermont.

K. BETH O'DONNELL earned her master's in regional planning from the University of Massachusetts at Amherst in 1997. As a municipal planner with a con-

309

sulting firm in Westchester County, she works with New York–area communities to implement sound planning practices and environmental review procedures.

RUTHERFORD H. PLATT is a professor of geography and planning law at the University of Massachusetts at Amherst. He served as the principal investigator of the study, supported by the National Science Foundation, that led to this book. He holds a J.D. and a Ph.D. (geography) from the University of Chicago and is a member of the Illinois bar. Among many publications, he is the author of *Land Use and Society: Geography, Law, and Public Policy* (Island Press, 1996).

CLAIRE B. RUBIN is a social scientist who is an independent consultant in the field of emergency management. She heads her own firm, Claire B. Rubin & Associates, in Arlington, Virginia. As a consultant, she has worked for many organizations—private, nonprofit, and governmental—during the past 20 years. Her work includes basic and applied research, the development and presentation of training programs, and the creation and operation of various information dissemination and utilization efforts.

DAVID SCHERF is an environmental educator at the Woodruff J. English Environmental Education Center in Claryville, New York. He currently serves as the center's watershed educator and is developing the Roehm Technology Learning Center.